LITTLE DEADLY THINGS

LITTLE DEADLY THINGS

A NOVEL

To Jo & Steve —
Remember — It's
the LITTLE THINGS
that count!
Best,
Harry

HARRY STEINMAN

Alloy Press

LITTLE DEADLY THINGS

Author photo: Barry Schneier
Cover and design: Roger Gefvert

LIBRARY OF CONGRESS CATALOGING-IN-PUBLICATION DATA

Steinman, Harry.
    Little deadly things / Harry Steinman.
    p. cm.
    LCCN 2012945815
    ISBN 978-1-9389526-0-9
    ISBN 978-1-9389526-1-6
    ISBN 978-1-9389526-2-3
        1. Women scientists—Fiction. 2. Nanotechnology—
    Fiction. 3. Medicinal plants—Fiction. 4. Science fiction.
    5. Suspense fiction. I. Title.

    PS3619.T47642L58 2012   813'.6
    QBI12-600160

Published by Alloy Press. For information, address:
Alloy Press
29 Prospect Avenue
Winthrop MA 02152
info@alloypress.com
www.alloypress.com

Library of Congress Control Number: 2012945815

ISBN: 978-1-938952-60-9

10 9 8 7 6 5 4 3 2 1

To my father, Jody, Rory and Rachel—the past, present, future and future perfect generations

In the hierarchy of Taíno deities, Yocahu was the supreme Creator. He lived in the northeast mountains, in the rainforest called El Yunque.

Juricán was the god of evil and the hurricane. He was perpetually angry and often turned on his own followers.

CONTENTS

PART ONE

BEGINNINGS

"WHOSO SHALL OFFEND ONE OF

THESE LITTLE ONES WHICH BELIEVE IN ME, IT

WOULD BE BETTER FOR HIM THAT A MILLSTONE

WERE HANGED AROUND HIS NECK, AND THAT HE

WERE DROWNED IN THE DEPTH OF THE SEA."

—Matthew 18:6

FEASIBLE CONTROL

BOSTON, MASSACHUSETTS

WEDNESDAY, MAY 19, 2038

2:00 PM

Eva Rozen strode into the waiting room looking like a shrunken wraith—girlish, ghoulish.

The 29-year-old scientist and entrepreneur had pale skin that could be compared to alabaster, if one were to be charitable, like plaster if not. It was pulled taut against the uneven planes of her face to produce an impression of constant tension, of perpetual threat. To look at her for more than a few moments was to falter, to lose one's balance.

This tidal wave in human form could move with great stealth but today Eva Rozen surged up to the receptionist's station trailing disturbance like a gunboat's wake. Sparks flew up from the clinic's marble floor where her heels struck, and the air boiled around her.

The office administrator looked up and froze. She'd been linked, but stopped speaking midsentence and tapped a small skin-toned communication patch just above her jaw to end the conversation.

Eva held out her hand to the attendant. The gesture was not an old-fashioned handshake. She dismissed ordinary social actions, and especially any that required physical contact. Rather, the act was part of a communications protocol. It signaled that Eva was using her datasleeve to gather the receptionist's cloud data, her public information: Bethany Jamison, genetic female, age 41, no criminal record. Eva's sleeve displayed all manner of private information as well. Jamison's credit profile, medical history, sexual preferences, augmentations, and other private and presumably secure data were available to Eva at a glance.

Armed with the administrator's name, Eva demanded, "Bethany, get Jim Ecco." Bethany Jamison, genetic female, age 41, no criminal record, did not move.

"Bethany," Eva repeated, "get Ecco. Now. Tell him Eva is here."

The administrator struggled to regain her composure. "Uh, I don't, that is, he's with a resident," she managed, "and *Mr.* Ecco has a full schedule," recovering.

"His residents stink. Tell *Mr.* Ecco that *Dr.* Rozen is waiting." Then, an afterthought, "Please."

Rozen's glare activated Bethany's survival instincts. The unflappable gatekeeper of Boston's largest animal shelter jumped up and scurried down a well-lit hallway. Two minutes later Bethany reappeared, stopping well short of the reception area. She looked once more at the visitor and then dove into an examining room like a soldier seeking cover.

In her place stood James Bradley Ecco—behaviorist, trainer, and chief handler of Haven Memorial Animal Shelter's three-score rescued dogs. Eva nodded a brief greeting that took in her old friend. He smelled of musk with traces of ammonia. Stray hairs left multicolored streaks on his uniform—tan scrubs with a dark blue logo, a paw print, and the word 'Haven' embedded over his left breast. His name, employee ID number, photo, and title glowed beneath the logo. His slight frame gave an impression of insubstantiality that belied his strength, speed, and anger.

The dog trainer, husband, and father could claim another distinction: in the entire world, he was Eva Rozen's only remaining friend.

Jim's face lit up. "Eva! This is a surprise. What are you doing here? I mean, it's great you're here. What have you been—"

He stopped midsentence. Few would have noticed the tightening below her eyebrows and fewer still would have recognized Eva's sudden impatience. Jim Ecco seldom missed small warnings, neither in dogs nor in people. He adopted a relaxed posture and leaned back imperceptibly, giving Eva an inch more space to signal his respect for her.

"Talk to me," Jim said. "What exciting plans do you have? Decided to adopt a puppy?"

"Nope. Don't need any research animals."

One corner of Jim's mouth turned up in a half grin. "Ah, Eva," he sighed theatrically. "Ever the humanitarian."

"Jim Ecco, ever the idealist. You make any kind of a decent wage cleaning up after dogs? Or does Plant Lady carry you?"

He ignored the barb. "Here to sell me stock in NMech?"

"Not going public just yet. But you should help me. Leased medical nanoagents means—

"Means NMech grows rich. Where on earth did you get the idea of leasing medicine, anyway?"

"I copied my strategy from King Gillette."

"King who?" asked Jim.

"King Gillette. That was his name, not a title. He invented disposable safety razor blades, figuring that he could just about give away the razor but charge for the blades, as long as they were short-lived. He made a fortune. It'll work for medicine, too."

"Rent-a-remedy?"

"Meds for the masses." Eva's eyes tightened again. Jim turned just a degree or two. It was the same indirect body posture that he would adopt with an agitated dog in his care. He fixed his gaze just to the side of his old friend and then looked down at the floor. Eva relaxed.

"I need your help," she said abruptly.

"My help? Or Marta's?"

"Yours. Hers. Both."

"We've been through that," said Jim. "We made a family decision."

"Yeah, well, I can sweeten the deal. I have something that will interest the Plant Lady."

"Oh? What's that?"

"Take my word for it. She'll like what I can put on the table. First I have to know if she can give me what I want." Eva, the Needy.

"First, I need to see my patients. You know, the stinky ones? Then we can talk."

"Patients? The vets have patients. You have shovels." Eva, the Relentless.

Jim did not respond. She would take any response, even correction, as tacit approval.

"I'll wait," she said. Eva, the Unexpected.

Jim stared. "You'll wait? Now I'm confused. You stormed in, a woman on fire. You lit up poor Bethany, demanded to see me, and now you'll wait? Why didn't you just link to me? I'd have been expecting you."

"I need to talk to you, that's why."

"Yes, Eva, but most people link ahead. Courtesy doesn't take that much work."

"Overrated," she snapped. "When it's time to do something, it's time to do it. Besides, I checked your schedule and I knew you'd be here."

"You checked my...?" Jim looked down at his datasleeve and frowned. "You're still up to your old tricks."

When Eva said nothing, Jim conceded, "Okay, it must be really important. Make yourself at home." He smiled and walked back to the kennels.

Eva stood still in her friend's cramped office. Only her eyes moved as she examined her surroundings. After a few silent minutes, she frowned and ran stubby fingers across her scalp, leaving rows of dirty blond hair like freshly-plowed farmland.

She took in Jim's neat piles of old-fashioned books and dataplaques. They were stacked on every available surface except Jim's small desk, bare save for a coffee service. Eva touched her datasleeve and launched a snooper application to search for key words, terms, and algorithms on the chance that any of Marta's work would be on Jim's datapillar. In the time that it took the handler to complete his morning duties, Eva sifted through Jim's pillar in an unsuccessful search. Her frown was a brief departure from her normal expressionless demeanor.

When Jim reappeared, he was covered in dog hair. Leashes hung around his neck like leather boas. He looked drained. He shuffled to his desk and brewed coffee. "Ah, sweet elixir of life," he said with his first sip and offered a cup to Eva who shook her head impatiently.

"Anybody ever tell you that you're full of crap?" Eva asked without rancor.

"Only you, darling."

Eva stood very still. She cocked her head, heard sounds within her that grew and threatened to drown out the sounds around her. She blinked hard, forcing a moment's quiet, and when she could hear again, she said, "Now it's darling?"

"Eva, you could charm the skin off a snake. Wouldn't you rather have a lifetime friend?"

Eva said nothing. The inner din quieted.

Jim gave no sign that he'd observed her distress. He lifted his cup in a toast to his old friend. "You never do anything without a purpose. And to wait without complaint? What's up?"

Eva looked at Jim. "I've got a problem. My project at Harvard produced two medicines—"

"Your project?—"

"—Okay, our project at Harvard produced two medicines that Plant Lady extracted from her rainforest. We showed that they could be built in a nanoassembler. I turned that project into NMech. I damned near went broke building an assembler that could be implanted in a patient's body. Getting FDA approvals was murder. But we're making a little money now."

"You've got working internal assemblers?"

"Didn't I just say so?"

The old manufacturing paradigm, whether for medicine or metals, was to whittle larger hunks of material into smaller ones. Nanotechnology, the science of matter at a scale so small as to be nearly unimaginable, permitted products to be built up, molecule by molecule. Eva's company focused on synthesizing drugs using this technology.

"Congratulations" Jim said. "So, what's the problem?"

Eva said, "Control. Let's say that you implant a nanoscale assembler inside a patient's body. It fabricates and dispenses the meds automatically. But you need to be able to raise or lower the dosage to match the patient's condition. We need to control the assembler after it's been implanted."

"An assembler at sub-cutaneous scale? That's science fiction, stuff of the future. What am I missing?"

"Everyone seems to think that progress is fantasy—until they get their nose rubbed in it. About one-hundred years ago, movie producer Darryl Zanuck predicted the end of television. Twenty years later? The world watched the first moon walk on their televisions. Trust me, I know exactly how to build a nanoassembler, and make it small enough to be implanted. What I need is control."

"Control is old hat," said Jim. "Doctors have had wireless control over drug implants for years."

Eva interrupted. "The problem with wireless control is the physician. Too busy to keep track. And the patients? They're worse. They skip appointments to recalibrate the implants. But if I could control nanoagents remotely? A Boston administrator manages a Berlin patient's prescription? If I can control the dosage from a datapillar, I can make the system efficient."

"You mean, you can control the cure."

"Don't get high-and-mighty with me. The system now means that physicians have to do a technician's job. Not fair to the doctor, not fair to the patient. But NMech has the medical skill and the technological know-how to manage the dosage remotely."

"Eva," Jim said gently, "she wants to practice medicine, not be a drug manufacturer. You know that."

"Let me finish. I have a proposal that she will like, if remote control is feasible."

Jim said nothing for a few moments, then subvocalized. His lips moved but he pronounced his words silently. A skin-toned comm-patch seated on his cheek registered the minute vibrations in his jaw and throat from the silent speech and converted the resonance into electronic pulses, and then sent a series of commands to his datasleeve. The sleeve activated a heads-up holographic display. Jim peered into the projection for a few minutes.

"According to Marta, the theory is simple enough."

"You can access her notes?" Eva asked. She held up a hand like an old-time traffic cop. The gesture indicated that Eva wanted to receive a datafile, the information that Jim had examined.

"No. These are her files. You want them? You ask her. If she wants to share, then she'll share." Jim spoke in a flat voice, a momentary withdrawal of camaraderie.

"Fine," said Eva. "Tell me how you get excorporeal control of a nanoagent. And, yeah, I will take a cup of your magic coffee." Jim touched a pot. In a moment the pot glowed gently and Jim poured hot water into a French press and set a cup in front of her.

"It's already been done, just not well," Jim said, peering back into the display. "About twenty years ago, researchers at Chambers Hospital implanted a tiny reservoir under the patient's skin to dispense medication subcutaneously. They added magnetic ferrite nanoparticles"—bits of material measured in billionths of a meter, near-atomic size—"to the reservoir. When the researchers turned on a magnetic field, the reservoir's membranes heated, and then turned porous. That released the medication."

"I know about the Chambers trials," Eva waved dismissively. Her coffee sat untouched. "But the membranes overheated and dumped the entire reservoir into the bloodstream. That's not going to work."

"True, but the idea is still good. At Chambers, they used a steady magnetic pulse which caused the overheating. What you need is a reliable regulator, and Marta thinks it's possible with something called chameleon magnets. Take a nonmagnetic material and hit it with an electronic pulse to organize the spin of the electrons. That turns the material magnetic. The reservoir's membrane heats and turns porous and delivers the medication. Turn the field off, the dosage stops. That would make an effective regulator. But the research on chameleon magnets was in the field of computer science, not in medicine, and nobody ever put the two ideas together."

"Huh," said Eva. She thought a moment. "Could the control signal come from a central datapillar? And be relayed to a home pillar?"

"I don't see why not. But that's not my area. You want to talk nanomeds? Ask Marta."

"You mean it? Plant Lady can do this?"

"She could, but I don't see her abandoning her work." The two sat in an amiable stalemate. "And by the way, it shows respect when you use her name. Marta. Not Plant Lady."

Eva fidgeted but did not speak. They'd discussed respect and social graces often, especially since the fiasco at Harvard. Eva's impulsiveness had cost Marta's trust and friendship. Jim counseled Eva that to temper sudden actions, to use proper names, to be courteous, even to observe ordinary table manners, were better ways to recruit help from others.

Finally, Jim broke the silence. "You need her, don't you."

Eva said, "Is that supposed to be a question?"

"You'll have to come up with something in public health," he said.

"Easy. I have a plan."

"Your last big idea was to build fat-loving nanomolecules for tummy tucks and replicator cells for breast augmentation."

"Boob jobs pay."

"Yeah, but you're not going to get Marta interested unless you go deeper into the thoracic cavity. She wants public health, not private wealth. You want her help? Meet her halfway."

"But you know what my problem is," said Eva.

"Sure. Chronic disease is expensive. And the countries that need help the most don't have the treasury to pay for it. So, you're back to where you started."

"Not this time." Eva's coffee was untouched and cold. She looked down and said, "If Plant Lady gives me the meds and the controls, I'll give her public health."

Jim stared without speaking.

"Sorry. If *Marta* can come up with the controls, then I'll give the good Dr. Cruz her public health."

"Do you mean it?"

"Yes," Eva said. "But you need to convince her. I can't just drop back into her life. And I need your help, too. You have a practical side that'll be valuable to NMech."

"Thanks, but I'm pretty happy with my stinky residents."

"Just hear me out," said Eva. "You and Marta will like what I have. Then you decide if my proposal is as important as your mangy dogs. Besides, you owe me."

"I haven't forgotten," Jim said. His eyes dropped to the floor again.

She had asked for something from him once before, something very personal, reminding him of his debt to her. "It's not mine to share," was all he said.

Then Mama and the others at the Table howled at Eva.

01

SUFFER THE LITTLE CHILDREN

SOFIA, BULGARIA

APRIL, 2022

One week before her departure from Sofia to attend a special high school in Los Angeles, 13-year-old Eva Rozen had awoken to the sounds of Mama and Papa fighting. She had been accustomed to shouted curses, taunts, and screams, even the crisp crack of hands on flesh. Those sounds had not bothered her. To be roused from sleep, however, was to lose its comforting amnesia. That did bother her, and a reckoning had been long overdue.

She slipped past Gergana's empty bedroom, gaze fixed ahead, and crept down the spine of the railroad flat to the fracas in the kitchen. Separate rooms and separate lives were connected by a dark hallway as grim as Eva's thoughts.

Eva stepped in unnoticed. Mama's screams alternated with

Papa's. Eva looked around. She heard a thought, as if from a separate intelligence within her. *Use what's at hand.* She found a wine bottle. It was an easy task. They littered the kitchen. She hefted one to test its weight, and gripped the neck, entered the field of combat and swung two-handed.

Eva was smaller than an average child on the cusp of adolescence and her aim was low. But she wielded the bottle with the predatory ferocity of a weasel and the roundhouse blow drove into Papa's left knee with a satisfying crunch. He bellowed as the kneecap shattered. Eva regarded her mother, swung and caught Mama just below her hip. The impact was cushioned by the soft tissue of Mama's thigh, once seductive territory that had first captured, then repelled, Papa. Mama cradled her leg, and sobbed. Eva regarded her parents, sprawled on the floor.

"That's for Gergana." Her voice was impassive.

She returned to her small bedroom where memories came, unbidden: Mama's indifference, Papa's drunken visits, and Gergana. Most of all, Gergana. Eva imagined what Gergana might have said to her tonight, tried to feel Gergana's cool hand on her forehead. She would have told Eva that she was very brave.

She wanted to sob but choked back her tears. At that, she heard a low murmur of approval. Startled, she sat up and looked about. The whispers would not have been from Papa or even Mama. They were still in the kitchen. Papa was moaning in pain and begging Mama to call an ambulance.

Eva heard the murmuring again. It was distant, yet…interior. For a moment, Eva imagined the voices coming from within her pillow. She sat up and then walked to the door. The sound grew and followed her. She heard notes of pride, of encouragement, of approval. The words were indistinct yet the meaning was clear: she had done well.

Then a second message emerged from the swelling clamor, increasing in volume, building in pitch and resonance, blotting out any other thought, a boulder rolling slowly at first, then crushing every obstacle in its path. "Strike first!" she heard from within the din. "Strike hard!"

Eva listened. She could pick out individual calls and caws. She strained to identify one voice. It would have been a quiet one. But Gergana's crooning was lost in the uproar.

❀   ❀   ❀

From Eva's first hours of life, uneasy forces shaped her. Her birth was a brief cause for celebration, as Gergana's had been some years earlier. Mama and Papa displayed Eva like a gauche traveler waiving a first-class airline ticket. But soon, parenting enervated rather than enlivened them and Mama and Papa's interest decayed. Eva was demoted from an object of inestimable worth to that of a curious gewgaw. Then they nurtured Eva as they might a caged falcon, tossing scraps of attention as they might have cast bits of offal to the raptor. The bird survives but is stunted, fettered by self-doubt, never to soar, always ready for a sharp-beaked defense of its circumscribed territory.

The roles of mother and of father fell to Gergana whose *de facto* parenting was as tender as Mama and Papa's was feckless. When Eva looked for comfort, her eyes lit on Gergana's smile. Eva's ears heard her sister's soft lullabies and her hands played with toys that Gergana somehow provided. When Mama ignored Eva's cries, Gergana cleaned and changed her infant sister. When Papa stumbled home, Gergana stood at juvenile Eva's doorway.

Eva nursed on Gergana's attention and Eva's loyalty was as fierce as a samurai's. Gergana adorned their bleak lives with bedtime stories, fanciful embellishments to bring hope.

"Little One," she'd say, "I've got a story for you." She portrayed the family as heroic figures in a romantic adventure. Gergana transformed Papa into a sea captain whose perilous journeys accounted for frequent absences. Mama was a member of the exiled royal family of Simeon II. "Little One," Gergana told Eva, "one day you'll be on that throne."

Gergana's stories and dreams were grand, but Eva saw life with open eyes. She fought to reconcile Mama's weak chin and perpetual air of distraction with the royal station of Gergana's tales. Eva saw Papa return, not from the high seas with sun-bleached hair and the tang of brine, but from a nearby tavern, red-faced, stinking of tobacco and stale beer. Sometimes the stench lingered, and Eva showered before returning to sleep.

"I don't want a story tonight," Eva announced one evening. Gergana had seen to Eva's bath, changed her into nightclothes and shut Eva's bedroom door so that Mama's weepy ramblings and Papa's snores were dampened.

"No story, Little One? How about a song?"

"No."

"How come? You like bedtime stories."

"They're not true. You made them all up." Eva's voice was flat, almost uninterested.

"They're supposed to be made up. Something nice to think about before you go to sleep."

"Mama's not a princess. Some days she doesn't even get out of bed. Papa is no sea captain. Sailors have sunburns. Papa's skin is all white."

"Well, stories are for pretending. I don't have to make up stories. I could read books with fairytales," Gergana offered.

"No. I don't want stories anymore. That's for little kids," said nine-year-old Eva.

"Well, don't you play pretend games with your friends?"

"I don't play with the other kids. And I don't like pretend games."

"What about your friends? Don't they like to play house or have tea parties?"

"I don't know," Eva said. "Anyway, you're my only friend. And I'm not Little One anymore. I'm a woman."

Gergana chuckled. "You're a woman now? How very grown up. When did you become a woman?"

"A while ago," said Eva, in a matter-of-fact voice.

"Oh, a while ago, eh?" Gergana teased. "And how did you decide that you're now a woman?"

"Papa told me."

❁   ❁   ❁

Gergana began a vigil outside Eva's door when Papa staggered home. She was Eva's guard. Sometimes she was Eva's alternate. She had no choice. She would protect Eva, no matter what.

❁   ❁   ❁

What replaces fantasy and imagination for the child exiled from the acres of make-believe? Where does the mind travel when fairyland becomes forbidden territory? Eva found sanctuary in science with its logic and its immutable laws. Banished from enchantment, Eva found chemistry. She could create new worlds, real ones. Leave illusion to children who could pretend in safety. Science offered Eva the means to travel from her perilous world to an orderly one.

❁   ❁   ❁

Sisters grow and sisters change. Gergana ripened into eye-catching adolescent beauty. She bore the hallmarks of classic loveliness—symmetrical features, full lips, high cheekbones and captivating green eyes—and her interests centered on boys. Gergana's breasts were full, and she turned and stretched to display them. Her toned legs drew admiring eyes up to wide hips. The owners of those eyes sought to accompany Gergana. Eva no longer had an unrivaled claim to her sister's attention.

Eva considered herself in a mirror. Her hair was unkempt, her features mismatched. She had no experience with style. Her single experiment with makeup led to calamitous results.

"How come you're so pretty and I'm so ugly?" Eva asked one evening as she walked into her sister's room following Gergana's return from a social outing.

"Would you knock before you come into my room?" The tone was abrupt.

"Why are you ignoring me? Those boys don't care about you as much as I do."

"Well, I like boys and it gets me out of the house."

"I wish you would play more with me," said Eva.

"Little One, we're not little kids anymore. You're my sister and I love you. But I have friends. You will, too."

"I doubt it. I'm not pretty like you are."

Eva clung to her sister but she was as awkward as a skittering foal and her efforts to hold onto Gergana fed the distance between them. The distance grew as Gergana's experiments with boys became experiences with men, her delight in schnapps and then liquor broadened to include marijuana and then cocaine.

Late one night Gergana stumbled home. Her key fought with the lock until the tumblers clicked into place and she staggered in. Her hair was matted, her clothing rumpled. Her words were slurred and coated with the sweet aroma of a flight of vodka. Eva helped Gergana into her bedroom, helped her get undressed. All the while, Gergana was singing popular songs or talking about her boyfriends, comparing one to another.

"Why do you do this?" Eva asked.

Gergana was lying on her bed. She reached for a stuffed animal, a plush pink rabbit with a blue waistcoat. Thin flexible wiring inside the toy's ears held a shape, and Gergana alternated between bending the ears down, flopped over one moment and then alert and erect the next. She brought the bunny up to her face and cooed to it as she stroked its length.

"Why do you do this?" Eva repeated.

"Do what?"

"Get drunk. Get stoned. Give yourself to the boys. That. Why do you do it?"

"Flopsy," Gergana whispered to her rabbit, "Little One is jealous." Gergana's words trailed off in an alcoholic haze.

"The boys don't care about you. I do. You should spend your time with me."

Gergana snored.

Gergana's widening social interests claimed her. Now, when Papa's clumsy steps shook the stairs leading to the Rozens' third-floor apartment, the post outside Eva's door was abandoned. Eva felt helpless. Her father was a big man, and she was small.

An unexpected warning from a surprising source gave Eva a solution to her growing dilemma. It was an ordinary spring day. Eva was dressed in her usual navy blue gabardine cargo pants. These were hemmed to fit her four-foot frame, cinched with a functional black leather belt that matched heavy black boots. A dark green work shirt gave her the appearance of a dwarfish custodian, and Eva's trademark black cloak made her look like a walking toadstool.

Mama's shapeless form greeted Eva that day. She was staring through eyes that were partitioned from the rest of her face by dark circles of fatigue. Despair carved hollows into her face. She shuffled along, wrapped in a frayed bathrobe despite the hour.

Mama started to speak, and then stopped. Eva had removed cleaning supplies from a storage area under a rust-stained sink. She held a bottle of bleach in her right hand and one of ammonia in her left.

"What?" Eva asked.

Mama stood just beyond Eva's reach. "You might not want to mix bleach with ammonia," Mama suggested.

"Why not?"

"It makes a gas if you mix them."

"What gas?"

Three simple questions from this adolescent girl carried the force of a State Security interrogation.

"Um, bleach has chlorine in it." She pointed to a label on the bottle. "See, 'chlorine bleach'. If you mix it with ammonia, it makes chlorine gas which can hurt you. What are you trying to clean?"

Eva showed Mama the offending spot. Mama examined the stain on the heavy fabric's sleeve. She reached for laundry soap and peroxide.

"Blood?"

Eva nodded. She offered no explanation nor did Mama ask for one. Mama dabbed the spot with peroxide, waited, and then scrubbed with laundry soap. She handed the cleaned shirt back to Eva and retreated down the dark hallway.

The label's warning puzzled Eva. It told her what was hazardous but not why. She went back to her room to find an answer on her bookshelf. Every volume was a text. Each bore the bookplate of Sofia's public library or of its university. Eva enjoyed the feel of paper and the heft of books and she handled them with the reverence of a rabbi cradling the Torah. Textbooks were her playmates and chemistry was her best friend.

Eva found her answer. Bleach breaks down in ammonia to release chlorine gas, a powerful greenish-yellow poison. During a world war in an earlier century, the gas earned the title of civilization's first weapon of mass destruction. Eva pondered these facts and determined that she must experiment and learn. It would be easy to find a subject for her investigation. Wild dogs roamed the streets of Sofia. Victims were widespread: the deer enclosure at the city's zoo suffered an attack and only the largest-antlered bucks survived; a British tourist in Nedyalsko, mauled almost beyond recognition; a child visiting her grandmother, dead. Eva had been set upon as well. She'd had the presence of mind to snap open an umbrella into the dog's face. Startled, the dog ran.

But dogs were fast and unpredictable. She needed a subject that she could anticipate and even control. She pondered this challenge for several days then found an answer, right at home. Why not Papa? Given his drunken gait up the stairs and off-key singing, she'd have enough time to prepare.

She found a squeeze bottle, one that would cap tightly and fit into an inner pocket in her cloak. Next, the formula. Equal parts ammonia and bleach would create a gas that would happily rip apart the delicate lining of a human's respiratory system.

Eva thought, *I'm on the right track, but not with chlorine gas. It'll get me, too.* There was another option, a favorite tool of the police and military—pepper spray. The recipe was simple. The active ingredients, the ones that burn—capsaicin and dihydrocapsaicin—came straight from hot peppers. The hottest of the many peppers available at Sofia's markets was Guntar chilies, imported from India's Spice Coast. This variety possessed commanding levels of the hot capsaicin molecules. The pepper's oily juice was soluble in oil, mineral oil for example. It would stick to its target and she could use it at close quarters. Or better still, baby oil—the scent reminded her of Gergana.

Eva's pepper spray ended Papa's nocturnal visits. The blisters around his eyes and mouth lasted for three days. His hunched-over walk lasted longer. After two encounters he was conditioned to avoid her door.

But he was like a puppy, and had occasional lapses. Eva learned to stay alert. Soon she had the reflexes of a combat soldier and could come out of sleep in an instant, ready to protect herself. Once Papa moved more deliberately. He caught Eva by surprise and she suffered the effects of the pepper spray along with Papa. Soon, her arsenal included a knife.

Papa was thwarted. He turned his attention to Gergana who found safety by increasing her the distance from Papa. That meant more time away from home at a time when Eva needed a big sister all the more.

How could Eva find a way to bring Gergana back to her, to regain the warmth of their earlier years? She decided to appeal to her sister's vanity. *I'll find a pretty present. Something for her to wear. She'll look in the mirror and think of me.*

Eva hunted in Sofia's stores. She sought just the right talisman to rekindle the magic that once existed between the sisters. A week into her quest, fresh from a successful raid at a library branch and clutching a chemistry text thick enough to double for building material, Eva spotted a small brooch in the window of an antiques store. It was a gold beetle studded with red, blue, and green stones. *It's pretty, but it's odd*, she thought. It was a scarab, an insect that feeds on feces.

Eva walked into the store. A pleasant musty smell announced a different world. Crowded displays of curios and keepsakes, used furniture and costumes from bygone eras greeted her. A squat man with a gray-speckled beard sat hunched on a stool behind the counter. The shopkeeper looked up from a leather-bound book. His black fedora, fashionable sixty years earlier, topped a bald head ringed with a graying fringe. A necktie from the same era—a wide silk paisley with what Eva thought were impossibly cheerful colors—wrapped his pale neck. He looked up from his reading, registered Eva's presence with a gracious nod. He moved deliberately as he marked his place with a well-worn leather bookmark and presented a pleasant expression. Eva found herself drawn to the odd-looking man. His eyes narrowed, not with suspicion, but because his wide smile lifted generous cheeks and his eyes had no place to retreat from the spreading grin.

Eva stood motionless, waiting for him to speak.

"What are you reading, young lady?" he asked, pointing to the thick text Eva clutched in her left hand.

"It's a chemistry book."

"A serious topic. And on paper, as well. Do you enjoy it?"

"Chemistry or paper?" Eva asked.

"Well, both, I suppose. It is a bit unusual to see anyone carrying paper except collectors. Why not read on your dataslate?" His voice was soft, restrained even.

"I like paper. I can get these out of the library without a fuss." She didn't add that "fuss" meant returning the books. "What about you? What're you reading?"

"Ah. A treasure. A first edition of *Alice in Wonderland*. Would you like to see it? Have you read Alice?" He turned the book around to face Eva.

"No. I don't read made-up stories."

"Not at all?"

"No." Eva's voice was simultaneously flat and emphatic.

"What a shame! There is a whole world of imagination that's waiting for you." The shopkeeper put his book away and muttered, "No fiction. What a shame." He turned back to Eva, "What about history or poetry? Art?"

Eva frowned.

"Just chemistry? You must love books to carry one so heavy from the library. What else do you read?"

"Science books. Computer texts."

"That's it?" The merchant drew up his eyebrows.

"That's all I need."

"What about the classics? Studies of the human soul?"

Eva said nothing.

"Well, my dear, if you ever wish to dip your toe into the ocean of human experience, come back to my little shop. Maybe we can find something enjoyable for a, a *serious* reader. You may well find that the study of people will make you a better scientist. Now. To what do I owe the undiluted pleasure of your company?"

She frowned at his flowery speech and pointed to the pin that caught her attention. "Why would anyone want a shit-eating bug for jewelry?" she asked.

"You recognize *Scarabaeoidea*? Good for you!"

She waited for him to continue, thinking that he would make a better teacher than the drones at school. This man seemed to invite her into his world, not try to force her. He was as different from her teachers as a guide from a kidnapper.

The shopkeeper's smile continued to hold court on the man's face. "Why indeed. Well, the ancient Egyptians held these little insects in high regard. The scarab was a symbol of rebirth."

Rebirth? Perfect. A gift to renew her relationship with Gergana. But the pin was rare and valuable, the shopkeeper said. He named a price. Eva thought for a moment and made a counteroffer. "Give me the pin and I'll give you protection."

He laughed, a hearty sound that rose from a deep well of joy. "Protection from what? Why would I need protection and how would a chemistry student provide this wondrous service?"

"Dogs. I'll keep them away. I'll keep your sidewalk clean. No more dogshit."

The shopkeeper tipped back his fedora and rubbed his forehead. He walked around the counter, navigated between an antique wheelbarrow and a child's rocking-horse to the window display, his ample body surprisingly agile. He took the pin and returned to place it on a square of black velvet on the counter. The two stood side by side and examined it. Jewels from the brooch sparkled against the cloth's plush black surface.

"I'll tell you what, Miss Scientist. You clean my sidewalk for one month and I'll give you the brooch. Don't bother the dogs. You could get hurt and I do not wish to risk the loss of such a valuable new customer."

When the shopkeeper arrived to open his store the next morning, a faint odor of bleach replaced the smell of feces and the sidewalk shone. He stopped and perused the storefront. In a voice louder than one would use when talking to oneself he said, "Beautiful. Just beautiful." He waited a minute, then without turning, he called over his shoulder, "Welcome back, Miss Scientist. You did a very nice job. Of course, you know that because you heard me say so when I arrived. Would you care to join me for a cup of tea?"

"How did you know I was watching?" Eva asked.

"I heard you. I live in a quiet world. Why would you sneak up on me?"

"I didn't sneak. My world is quiet, too."

The proprietor held the shop door open for Eva. Today he'd replaced the fedora with a gray homburg, the wide brim turned up all the way around. A long white feather shot from the hatband, transforming the semiformal headwear into something jaunty. On another, the feather would be an affectation. But on this man, it was an antenna that transmitted his vitality.

The strange pair entered, a sedentary looking older man bristling with energy and a diminutive child brimming with strength. Without a word or backward glance, the man walked into a room behind the store. Eva followed into an office-cum-kitchen. The shelves were lined with books from earlier centuries. An antique red and blue Persian carpet muted their footsteps. Two parallel walls were painted yellow, as bright a color as she could imagine. Their counterparts were a correspondingly deep blue. A triptych of paintings hung along one of the long yellow walls, three masses of color, each swirling in a tight pattern of curves, streaks, and spatters. Along the back of the office was an antique partner's desk. Both sides were crammed with books and papers. To her right, Eva saw the shopkeeper at a small counter, fussing with a kettle and hotplate. A stained teapot, its glazing cracked with age, matched two antique white ceramic mugs. He whistled tunelessly as he worked, and gave Eva sidelong glances. When she caught him looking at her, he held her gaze and smiled. Tiny pastries appeared on a small plate and suddenly there were two chairs and space on the counter for their impromptu snack.

The store owner gestured with an outstretched hand to the tea and pastries and they began a snack and a silent conversation. Eva nodded acceptance of the invitation and then pointed with her chin to the trio of paintings. The man tilted his head down and formed an arched eyebrow question. She looked at the reproductions and shook her head. Then she spoke for the first time since entering the store.

"What's your name?"

"Coombs, at your service."

"Is that your first name or your last name?"

"Just Coombs. And you?"

"Eva. Coombs isn't a Slavic name. Are you British?"

"What do you think of the paintings?" he asked, without answering her question.

Eva looked at the framed art and asked, "Worms?"

He laughed. Once again, the sound was unforced. "I take it you're not familiar with the work of Jackson Pollock."

"You've got bugs for brooches and worms for paintings." She paused, considering, "I've never seen worms like that. The colors are wrong. It's not realistic."

"No, not realistic at all for worms. But Pollock didn't so much try to paint worms as he tried to make art without a brush coming between him and his creations. So he dripped paint on his canvases rather than brushing it on."

"You like these?"

"I do. Eva, look at them. If you wanted to make a painting of the energy in a chemical reaction, how would you do that?"

"I don't know. Not like that, I don't think."

"What about, say, Brownian motion?"

"These paintings are supposed to be the random movement of molecules?"

"Good. Now think bigger. Pollock was trying to show the energy and movement of life. That's my opinion, anyway."

"It looks like a baby's scribbling."

"Maybe yes, maybe no. Look deeper, Eva. What he did was to use things he could control—the thickness of the paint, the movement of his body, how absorbent his canvas was—to portray things he couldn't control. It looks chaotic, but isn't life chaotic? Don't we all try to control the chaos around us? That's what I see in his work. Think of chaos theory and then imagine it as art. You just might end up with Jackson Pollock."

"So what? Why would anybody want to paint science?"

"Art can inspire science."

Eva gave a snort.

Coombs continued, "A sculpture that looked like a tower of needles inspired a major breakthrough in understanding cell structure. Four hundred years ago, the divisions on a horsetail plant inspired John Napier to discover logarithms."

"I don't need art to do science."

"Okay." Then, "How's your tea?"

They sat without speaking for several minutes. Eva stood and explored Coombs's work area and looked at his book titles.

"May I offer a suggestion, young lady?"

"Eva."

"Yes, indeed. Well, Eva, I have a suggestion. Your work cleaning the sidewalk was better than I expected. I should be taking advantage of you by offering only the brooch as full payment for this good a job. I'd like to give you a book, real paper, an old edition with some value."

"What book?"

"It's called *To Kill a Mockingbird*."

"How hard can that be?"

"Eva, it's not a textbook."

"Then what is it?"

"It's the story of a young woman like you. A good girl named Scout must face terrible things and terrible people. She has to struggle to be herself despite awful events that happen around her. I rather think you might enjoy reading about how she managed."

"How old is Scout?"

"When the book starts, she's five."

"I'm thirteen."

"You were five once, yes? And now you're older?" Eva nodded. "Well Scout grows older, too." Coombs went to his book collection and muttered, "I know it's here."

Eva continued to wander about the work area. She stopped at the Pollock triptych for several minutes. She said, "It's funny. I don't like stories because they try to tell you something is true when it's not. This—" she nodded to the grouping, "—doesn't try to lie. It doesn't try to pretend to be a picture of something. It might be nonsense, but at least it's honest nonsense."

"How does it make you feel?" Coombs asked.

There was a long pause and Eva turned away. She turned back to Coombs and said, "I have to go. Thanks for the tea."

"What about the book?" He reached back to the shelf for the slim volume.

But when he turned back, Eva was gone.

She did not miss her duty once, not even Sundays. Thirty-one days after first meeting Coombs, she skipped home, bobbing under her mantle, the brooch in her pocket. What a splendid gift she would present to Gergana. Eva imagined all of the things that they would do together, once again, and she smiled.

Eva's smile died the moment she crossed the worn threshold into the Rozen apartment. She heard hoarse cries of pain from Gergana's room, exhausted pleas in place of Gergana's insubstantial chatter.

Eva edged to the door, paused, listened and heard the crack of a palm striking flesh. There was a muffled thud followed by an explosive whoosh of air forced from unprotected lungs. Why today, of all days? When she had the brooch that would bring them back together and restore the magic they once shared?

She turned the doorknob, paused, and slipped into the bedroom. Her senses recoiled at the tableau before her. She registered the sour stink of sweat and hatred. An obese man was the source. He was naked, with blemish-mottled pallid skin. He lay between Gergana's legs with his hands loosely at her throat. Skin puffed out from his neck to give the impression of a bleached bullfrog. His face was frozen in a rictus, a grotesque parody of ecstasy.

Eva tore her eyes from the fat man and took in every detail in the room. The markers of Gergana's youth—stuffed animals and movie posters—were torn or trampled. She saw a broad-shouldered man in one corner of the room. His mouth was a compressed red slash. His bare chest was decorated with a heavy gold chain and thatched with a dense mat of black hair. Shards of pale blue ice, shaped like human eyes, looked from his face and focused on Eva. They froze her in place.

Bare Chest looked down at an old-fashioned wristwatch, and then back to Eva. When he spoke, she felt the paralyzing cold again. "You come to join, little girl?" Bare Chest asked. "I get good money for you. Better than your cow of a sister. Come here."

Eva could not move. Bare Chest closed the distance between them with feral grace. One moment he was seated, the next he towered over her, a steel-gripped hand wrapped around her left wrist. She was too stunned even to flinch.

Gergana moaned. "Nooo…not her. You promised. Not her," she croaked.

Bare Chest laughed. "This ugly runt is like a doll. She will fetch good money." He looked down at Eva, "You want to feel nice like your sister, eh? I give you something to make you fly like the angels."

Eva looked at Gergana. There were puncture marks on the arms that had held Eva. The face that had looked at Eva with adoration was livid with bruises. The eyes that had cherished Eva were swollen. Eva tugged but Bare Chest kept his easy grip on her wrist.

"Hey, Doran," Bare Chest called to the fat man. "You want this ugly runt? I let you have her cheap."

Eva looked up as the man called Doran continued to piston his hips and tightened his grip on Gergana's neck. Her eyes bulged.

"Don't mark her face," Bare Chest snapped, "Now she's lost value and you have to pay more." Doran relaxed his grip.

Eva was suspended in terror. Her eyes darted about, desperate to find something she could understand. Torn posters and stuffed toys. Gergana. The fat man. None of it made sense.

Bare Chest reached in his trousers and freed himself. He forced her unresisting left hand down and wrapped the girl's small fingers around his sex.

Eva said nothing. She merely complied. The room around her started to collapse into a pinpoint and her reason was pulled toward a black hole of fear.

Again, Gergana tried to lift her head. Again, she gasped, "No. Not her. You promised."

Without taking his predatory gaze from Eva, Bare Chest hissed at Gergana. "Shut up."

Still, Eva said nothing. She was nearing the event horizon of terror. In a few seconds, she would be lost.

She shut out Gergana's cries of pain and walled off her own terror. *That's better. Nice and quiet,* she thought. Then she heard another sound, one from within, first a murmur, then a tumult. A babble, then a coherent Voice shouted to her, *"Fight!"*

But how? she whimpered in silence. The Voice said, *You're a scientist. Use what is at hand. How did you stop Papa?*

She looked at Gergana and then back to Bare Chest. Her eyes narrowed and her head moved in a slight double-take, as if the obvious solution to an intractable problem had presented itself. Then she nodded: a decision proposed, seconded, and approved by acclamation.

Now Eva spoke with a steady voice. "Nobody calls me a runt. I'm going to kill you."

"Oh, you think so?" Bare Chest snarled. "For that I will split you open." He looked back to the fat man and said, "Hey, Doran, I give you the runt when I'm done. No charge. A present from me."

Eva heard herself shout—or was it something within?—*Now!*

Eva's right hand, her free hand, moved unseen under her cape as Bare Chest spoke to the fat man. She slipped her small squeeze bottle out from an inside pocket. One-handed, she flipped the cap open. The action was practiced and smooth, thanks to Papa's nocturnal visits. She sprayed Bare Chest's blood-engorged penis. It would be three seconds before his sensitive skin reacted to Eva's homemade pepper solution. In that time, Eva reached up. In a motion perfected by her encounters with Papa, she sprayed Bare Chest's eyes. The effect was instant. His eyes reddened, bled, and bulged. Surprise, then agony, replaced his grin. Sightless crimson puddles replaced his ice-blue eyes. Then the oily fluid penetrated the dilated blood vessels in his penis. He screamed. Eva reached up to empty the remainder of the bottle's sap down Bare Chest's throat.

He gasped in agony. The pepper-laced oil coated absorbent tissues in his lungs, searing and choking him. His throat began to swell and his yelps of pain trailed off to a rasp. Eva, her right hand still hidden under her cape, dropped the spray bottle and drew her knife. She thrust the four-inch blade, aiming for Bare

Chest's genitals. She'd never used a knife in anger, and she missed her target. But the blade sliced neatly across his groin and severed the femoral artery. The resulting blood loss was instant and catastrophic. Bare Chest was still rubbing his eyes when he collapsed. Blood spurted as he bled out. A bitter stench filled the room as Bare Chest's bowels relaxed. Eva stabbed again. Her blade penetrated a corpse.

In the moment of Bare Chest's truncated scream, Doran lost all restraint. He bore down on Gergana. The weight of his body radiated through fat arms. Sausage-sized fingers dug into Gergana's slender neck. He pressed harder and now she convulsed, her feet moving in a frantic swimmer's kick, up and down, up and down. Then they were still. Doran grunted and relaxed in his release, Gergana relaxed in hers.

The fat man rolled off the body and towards Eva. She tried to ward him off, but she was drained. Killing Bare Chest had required that she reach deep into a core of rage and muster every bit of her strength. The simple act of lifting an arm was beyond her. Doran shot out one meaty fist and caught her in the temple.

Eva was sitting in a puddle of blood, wide-eyed and mute, when the police arrived eleven minutes after calls from several neighbors brought them up the three flights of stairs to the Rozen apartment. The responding officers found Mama asleep as they cleared the apartment. She could answer none of the questions posed by the police, nor could Papa, once he was located and escorted to the abattoir where Gergana once grew and dreamed.

"What happened?" one of the police asked Eva.

Eva was silent.

"Come on, miss. Something terrible happened here. Tell us what it was so we can find the man who killed your sister."

Eva was silent, still.

Mama came into the bedroom and screamed. "Tell them what you know!"

Eva looked up at her mother without speaking, and held Mama's gaze until the woman turned away. *I'll find him,* Eva said to herself. Then she tilted her head, as if listening to sound inside. The din was building at the Table.

Eva hunted for her sister's killer, relentless as a dingo. Her developing computer skills allowed her to glide through police files. She swallowed the coroner's documents, straining data, clinical observations and conjecture. The lurid crime scene photographs and the stilted clinical language of police reports were a grisly counterpoint to Eva's memories of the sister who had cared for her, who had offered tenderness when there was none from Mama and Papa.

Within the morbid affidavits, Eva found sketchy information about the fat man called Doran. For days, Eva saw his leering face and watched, again and again, as he strangled Gergana. His grunts echoed in her ears. She spent sleepless nights hunting and took on Mama's haunted look. Unlike Mama, Eva moved with purpose. She pursued her quarry with a vigor that would shame a detective on his first case. She sought prostitutes sporting bruises. She asked about Doran at every bar, café, and store in Sofia. One of these was Coombs's antique shop.

"Welcome back, Eva." Coombs was bareheaded today. He wore a concerned look. "How are you holding up?"

"You know?"

"It was in the news. You're a very brave young woman."

"I'm going to get him."

"I'm sure you will. But Eva, whatever you do, please—come back and see me."

Three weeks later, Doran's body was found in the Vladaya

River, under the Lion's Bridge. The skin on his face had a curious blistering, a unimportant detail to the detectives who were happy to close out three investigations—the rape and murder of Gergana Rozen, the assault of Eva Rozen, and the murder of Alexsandar Yorkuv—Bare Chest. They hung this crime on Doran as well. Who else in that apartment could have killed a grown man?

❀   ❀   ❀

Eva treated Mama and Papa with silent hostility. She broke her silence to indict, try, and convict them as accessories in Gergana's death. "You must pay," she ruled.

"We lose our daughter and you threaten us? What do you want?" Mama asked.

"Send me to America. I want to be away from you."

"Impossible. We have no money. We spent everything for your sister's funeral."

Eva stared at her mother and then handed her a brochure for a foreign studies program organized by the Hidden Scholar Foundation. The charity sought brilliant children from the most troubled neighborhoods around the world. These they gathered at magnet schools in the United States. Her school's principal had suggested the program.

"I will go here. You will submit my application."

Eva's stellar academic record bolstered the application as did an embarrassing wealth of recommendations from educators who dreaded her return. Fate smiled. The Hidden Scholar Foundation accepted Eva and provided a stipend for all of her expenses. She would enter a high school in America, in East Los Angeles.

Before she left Sofia, Eva kept a promise and returned to see to Coombs. Today he wore a round flat cap, its small brim snapped shut, and pulled low over one eye. Despite his girth and floral necktie, he looked mysterious, as if his hidden eye held a secret.

"You're leaving?" Coombs asked.

"How did you know?"

Coombs did not answer but beckoned Eva follow him, and the two walked back into his kitchen area. "Tea?"

She nodded and he directed her to a small three-legged stool near one of his bookshelves. There was a newly-framed composition on the wall facing the Pollock grouping. A woman looked out at the viewer. Midway down the work, the woman blended into another, upside down.

"This reminds me of a playing card," said Eva, "but the upside-down woman is different."

"How so?" asked Coombs.

"Well, either the artist isn't very good or he's trying to paint two different people. The one on the bottom has big arms and her face is, I don't know, sort of the same, sort of different."

"Could it be the same person, but two different ways of showing something about her?"

"I don't know. Their mouths are similar, and the shape of their heads. But one is so muscular and the other one is like a normal lady. It's like the same person on different days."

"Very good, Eva. The painting is called 'Portrait of My Sister and Picasso Figure.' The artist is Salvador Dalí. He painted things that he saw in dreams and his imagination."

"What's it supposed to mean? That his sister was weak and strong?" Eva asked.

"What's it mean to you?"

Eva frowned and studied the framed work. "This is a print?"

"Oh, my heavens, yes. What I wouldn't do to have an original Dalí. He is my favorite artist and his paintings are priceless. But, tell me, what does it mean to you?"

"I like that he could tell you two true stories about the same person. Maybe the woman has two different personalities."

Coombs nodded approval and said, "Tea's ready. I'll get some pastries. Meantime, grab the stool and see if you can find a book on the second shelf from the top. It has a light green binding. It's called *The Secret Garden*.

"Here it is," Eva said. "What's so good about it?"

"The heroine, Mary Lennox. She reminds me of you. Mary finds a healing garden. I think you need your own secret garden. Please, take the book. There'll come a time when it might make sense to you. Or if not that one, call me and I'll find you another."

Eva examined the book and then replaced it. "Maybe some other time."

"Please? Humor me. Take the book and read it someday. When you do, call me and tell me I'm a fool. Or ask me for another. I see great things for you, Eva, but I see struggles, too. The better you know yourself, the better you'll face your struggles."

"If I read it I'll let you know."

At that, Coombs smiled so broadly that his eyes all but disappeared. "My dear Eva, I shall hold you to that promise. And I should be honored if you would consider me a friend."

❀　　❀　　❀

Mama and Papa bade no farewell to Eva. Let this new land and its school have her. Mama had come to believe that her baby was a *hala*, the demon in Bulgarian lore who tries to consume the sun and moon and end the world.

By her thirty-sixth birthday, Eva would fully justify Mama's concerns.

02

THE ROZEN PLAN

Eva stared at Jim, waiting for an answer. She heard dogs barking in the kennels. Haven Memorial personnel came and went. Each glanced at her and then looked again, as if to confirm what their eyes had seen. Eva was accustomed to stares and ignored the attention. Anyway, getting Marta and Jim to work with her at NMech was the priority.

"I'll listen to your proposal, but no promises—even for an old friend. If Marta agrees, where would you start?" Jim asked. "Do you have a project in mind? Where are you going to find the capital to do the research, the manufacturing, and the trials?"

Eva waved away his questions. "All worked out. We can stay profitable and she can attack some of the issues that are important to her. I don't run a charity, but if she gives me what I need, then she can go help poor people all she wants."

"Do you mean it, Eva?"

"Ever known me to lie?"

"Not to me. What do you want?"

"Three things," said Eva. "First," she held up a stubby index finger. "Give me a control system for the nanomeds, something from a central source. Magnetism or magic, I don't care. But I need control."

"Let's say that's possible. What else?"

Eva held up a second finger, ticking off her list. "Help me introduce NMech's first health product."

Jim looked puzzled. "Medicine is a long-term deal. You've got simulations and trials. That could take years."

"Trust me, it won't," Eva replied.

"How can you be so sure?"

"I have a plan."

Jim chuckled. "The famous Rozen Plan. What's the third thing?"

"I need your wife. She'll listen to you. Get her on board."

"You sure you want to work with her again?" Jim stared. "I thought that you and she—"

"That's history. Disregard it. I need her. She's got some weird juju. She takes a walk in the park, comes back with a cure for something. She started by looking for a remedy for her JRA and now she has the largest library of plant-based meds in the world. I know about the work she did in Floresta Amazonica and the Borneo-Mekong. She even has friends in the Dzanga-Sangha Park in the Congo."

"For two people who haven't spoken in years, you seem to be up on her career."

"She knows what plants have medicinal properties. I want that. If it takes public health to get Marta to grace us with her knowledge, then the masses will have their day."

Jim shrugged. "Eva, this might work. Where will you start? How will you fund the research and the trials?"

Eva held up her hand again, another communications protocol gesture. This time Jim mimicked the gesture. Eva's cue told Jim that she was about to transmit a file. Jim's cue indicated willingness to receive. Eva subvocalized the commands to her sleeve. It emitted a focused electronic burst. Jim's sleeve interrogated to confirm the nature, source, and safety of the transmission, then pinged acceptance.

Eva studied Jim as he peered into a holographic heads-up display that projected from his dataslate. His eyes tracked back and forth as they scanned. Corneal implants, a bit like contact lenses of a prior generation, allowed him to read the holographic text. His eyes widened and narrowed as they pored over the file. His brows pulled down—first puzzled, then annoyed, and then angry.

"Eva. What the hell is this? You call this public health?"

"No, you idiot. It's exactly what it looks like: a simple over-the-counter remedy to fix a medically unimportant problem that no one has addressed. We don't need FDA trials for this. Labeling? Public process? Panel review? Yes. But clinical trials? No. The active ingredients are already approved. Just read a little further and you'll see why I picked this to start."

Jim shook his head. "Eva, you know something? You can be a real pain in the ass."

She beamed. It was her habit to get the better of others lest they get the better of her. It wasn't easy with Jim, but she counted coup.

At first, Jim's face betrayed no expression as he read on, then he grinned and started to laugh. Eva stiffened. *He thinks this is a joke,* she thought. She flushed and turned to leave. Now even the quiet Voices were raucous. Mama shrieked in derision.

"No, Eva," Jim managed to get the words out. "Stop." He threw an arm around Eva's shoulders and gave a fraternal squeeze. Eva stiffened for a moment, then softened and leaned into Jim's half embrace.

"Eva, you're too much. This is great." Jim was still chuckling. "I underestimated you. You're two steps ahead of us, as usual. I'll talk to Plant Lady tonight."

❀   ❀   ❀

Marta Cruz was steeping herbs when Jim palmed open the front door of their Brookline apartment. The low-grade fever was back. Fatigue and pain pulled the muscles in her face tight. She rubbed young stinging nettle leaves on her skin to produce an irritation that brought blood to the surface and reduced the swelling. Then she sipped her tea: false garlic, cascarilla, and chinchona bark. She'd been taught the remedy by her grandmother, her *abuela*. The brew had little to recommend by way of taste, but it would ease the pain.

"Don't just stand there," she said. "Come kiss me."

Jim smiled and complied.

"*Te quiero*," they both murmured. I love you.

"You look tired," Jim said. "Hard day?"

"The usual," she said.

Marta had been stricken with JRA—juvenile rheumatoid arthritis—when she was nine. The autoimmune disease provoked swellings, fevers, and rashes. It held a vise grip over her knees, elbows, and hands and lent a slight S-curve to her spine. On the days when it was difficult to stand, she used the herbs Abuela showed her in El Yunque, the rainforest.

Jim embraced her again and the two stood silently, each

drawing strength and comfort from the other. "Don't just stand there. Kiss me again," Jim said, parroting his wife's command. He held the embrace and pressed his face into her hair and inhaled. Then he kissed her again.

"Well, big boy, you're in a good mood. Did you have a good day at work?"

"In fact, I did."

"Something special?"

"I have something exciting to tell you. But I want you to keep an open mind, okay?"

Marta stiffened in his arms. "Does this something have anything to do with Eva? Did she come to see you today? Friend in need?"

"Marta, please. Just listen."

"Every time you ask me to keep an open mind, it's about Eva. That woman is toxic. Did she say that you owed her?" Marta looked at him and shook her head. "You don't have to say a word. I can see it in your face. Well, you've paid your debt just by being her friend."

"She's changed. Just hear me out."

"Changed? I doubt it. The best predictor of future behavior is past behavior. And I don't trust her when she's around you."

"Marta—"

"Sorry, but that's how I feel."

"If it weren't for Eva—"

"I know that. But she's a thief and she's carrying a torch for my husband. You expect me to welcome her back with open arms?"

"I'm hoping that, finally, you will. What she did was wrong, but she was young."

The room went silent. Presently, Marta drew in a deep breath and exhaled slowly. She pinched her ear. "Okay. Tell me what she's up to. Tell me why I should ever work with her again." Her voice sounded resigned, but she stilled herself and listened.

He began, "You have to admit you two made a formidable team in college. Your work in biology, her work in chemistry, and her business skills? You did some good science."

"I'm not sure it makes up for the rest. What makes you so sure that Eva won't do the same thing all over again?"

"There's no guarantee," he conceded, "but Eva seems more mature than she was at Harvard. Maybe running a business helped her control herself."

"No, that was the problem. She wants to control everything, and I don't want to have to be looking over my shoulder again."

"Look, Eva is the most driven person we know. When she puts her mind to something, watch out. All we have to do is keep her pointing in the right direction."

Marta considered. "It's tempting. Like a jewel heist is tempting. Okay, what's her grand scheme this time? No promises. Just tell me."

As Jim started to explain, Marta thought back to Harvard, to Eva, and everything that threatened to take her away from her rainforests. *Oh, Abuela, things haven't gotten much clearer since our summer together. I wish you could tell me what I should do now...*

03

 ## TAÍNA

LOS ANGELES, CALIFORNIA

EL YUNQUE, PUERTO RICO

APRIL 2022

The night before she left for Puerto Rico—the day after the funeral—thirteen-year-old Marta Cruz asked her father about the old woman's prediction.

Rafael Cruz didn't seemed to hear her or had chosen not to answer. Marta thought he looked lost in his own kitchen, gazing without focus around the East Los Angeles apartment. Marta's awards, drawings, and report cards covered one wall. A montage of photographs of a tropical forest covered another. Father and daughter sat at a worn grouping in the tiny kitchen, a table and three straight-backed, caned chairs. One chair was empty.

Marta moved with grace despite her limp, open-faced despite her sorrow. A halo of glossy black curls framed her pale skin, a remnant of the Spanish conquistadores who mixed their blood with the conquered—her father's caramel-complected people of Mexico's northern mountains, her mother's broad-faced Taíno, the native people of Puerto Rico. A roll of the genetic dice and recessive traits from each bestowed Marta with fair-skinned beauty, a hint of Iberian bronze that would deepen in the sun. Delicate facial bones outlined sharp features. Her eyes, as dark as her father's, were permanently curious and gave the impression that everything she saw was new.

"When I met your mother she was every bit as beautiful as you. Her hair was glossy, black, and straight. She brushed it one hundred times every night." He gazed into his memories, then shook his head and turned back to the inescapable present. "That was before the cancer ate through her." His voice trailed off.

"Dad?" She waited a few moments. "Dad? What about Abuela?"

Marta wanted to get her father talking again. It had been three days since her mother's death, and he seemed paralyzed with incomprehension. He even seemed oblivious of her own grief. She thought that her future might as well have been buried with her mother.

Elena Cruz had been the daughter of one of the last of the bohique, a medicine woman of Puerto Rico's Taíno Indians. But Elena looked past the flowers and plants and remedies that grew in the island's rainforest. Television had shown her a different beauty, ersatz splendor, effortless wealth. A different sun shone in Los Angeles and the forest's profuse bloom was reduced to a florist's inventory. Dull smears of crimson replaced the sunset arcs of red, yellow, and violet—the Caribbean's palette. That was before the night sweats and pain.

"Dad, tell me about when you and Mom met. Tell me about when you were happy. Please, Dad, that's what I need to hear." *If I have to leave my mother's grave, I need something to hold onto.*

Rafael's face softened. He looked at her, and for a moment, Marta saw his eyes brighten.

"You know that you're every bit as beautiful as your mother? No, more beautiful even."

Marta felt her eyes moisten and wondered when her father would show his own tears. If he would only give in to his grief, then maybe he could see her pain.

"I met her when I was a busboy at a fancy restaurant for *los ricos*. I hated the white linens I folded before service and the white skin and white teeth of the people around me."

"Dad—"

"Sorry. But that's where I met your mother and it is where the story begins. She was the housekeeper for the owner of the restaurant. La señora's house was in Malibu, by the ocean."

"One night la señora asked your mother to help in the restaurant. It was her night off. One night every two weeks. But if la señora asked, then who was your mother to say no? I was told to drive her back to Malibu. It was a long drive—two hours!—and I wasn't being paid for the driving. I knew it wasn't your mother's fault, but I was angry and I didn't speak a word to her, not once during the entire trip."

"When we arrived in Malibu, your mother pointed to a driveway that climbed up a steep hill from the coast. The house stood up on stilts, balanced on a hillside above the ocean. It looked like a shoebox with legs."

"I asked her about this crazy house. She told me that the hillside turns to mud when it rains and the houses slide into the ocean unless they are on pilings. I wondered if these people were so rich that they could throw away their houses. The thought made me dizzy."

He lapsed into silence.

"Dad?"

"Yes?"

"I miss her, too."

"Oh, *mija*. My poor Marta." Rafael reached an arm around his daughter's shoulders. She shifted to wrap both her arms around him. Minutes passed as they clung to each other.

"I'll tell you what I remember about Abuela's prophesy and you can ask her for the rest. I don't know if it will help. She speaks in riddles." He thought for a bit and continued. "She's known all over the island. When a child is sick or an adult is injured, they find her. When a wife can't conceive, when a farmer loses his strength, they turn to her for one of her herbs. Even the other bohiques come to her for advice. She could cure anything. Anything except Elena."

Marta stifled a sob, even as her head lay still on her father's shoulder.

Rafael sighed and kissed her forehead.

"During our wedding ceremony, Abuela took a handful of herbs from an old leather pouch she carries around her neck. She put the dried leaves into a tin can along with a white-hot coal and placed it at our feet. Soon a sweet smoke enveloped us. I breathed it in and started to relax. I did not even feel the hands that lowered me to the ground."

"I felt your mother's presence more than I saw her. I saw our lives like two vines, braided strands. Hers was a rich, deep forest-green and mine was the dark color of good earth. Then we saw a new strand, a brilliant gold that outshone everything else. That was you, *hija*, that was you."

Rafael shifted to face Marta. He reached and cupped her face. She felt the rough skin of his palms on her smooth cheeks. She hugged him tighter. He wrapped his arms around her and held her to his chest. She felt like she might break any moment.

"The vision—at first it was a beautiful dream. I was elated. But then I saw that this shining golden thread was wrapped with a black fiber that would choke it. That is when the vision ended."

The shaking started gently. It built within Rafael like water simmering to a boil. For the first time since Elena died, he gave free rein to his grief and sobbed. Marta wondered how he would survive the summer without her. How would *she* survive? She'd be alone. Why did her mother make her promise? Why was it so important to leave her father and miss the last days of school? To miss the summer?

❀   ❀   ❀

Elena Cruz's dying wish was that Marta go to be with Abuela. Marta protested, "How can I leave Dad?" He needed someone, she said, while she thought, *Why am I being sent away?* But her mother was resolute. "Promise me!" she demanded in a hoarse whisper. "You must go to your abuela. She will help you grow. She will help your legs. Maybe you will learn something from her to help your father. He's strong and proud but he's so frightened. Please—go and be with Abuela."

The flight to San Juan was a day-long course in agony. Her legs twisted in the cramped seats and pain ground through them like slow-moving knives. Gnarled swellings throbbed in her ankles and knees. She had a flu-like fever and a light pink flush dusted her skin in a way that would never be mistaken for a healthy glow. Lines of worry carved a map of fear into her face. She tugged at her black curls and then tucked them behind delicate ears, again and again.

Abuela waited as Marta stumbled into the terminal. Even in the jostling chaos of the crowd, her grandmother stood alone, unperturbed, travelers flowing around her like water around a rock.

Marta was exhausted and was grateful to let the old woman take her arm and guide her through the airport.

Neither said a word along the trip to the northeast part of the island. Marta's breath hitched. *If I start to cry now, I don't think I could stop.*

The trip southeast to Abuela's home brought them to a quiet haven. They took a shared taxi, a *carro publico*, out of crowded San Juan to an area just outside of Fajardo, in the northeast. As dusk fell, they passed a glowing bay, lit from within by microscopic creatures. Marta was too tired to consider the natural wonder before her. Its cold, green glow looked to Marta like an entrance into a world beyond. A chorus of tiny tree frogs peeped a cheerful welcome to the unhappy girl. Marta stumbled behind Abuela along a path that seemed invisible until the old woman pointed the way. She barely noticed Abuela's cabin as the old woman helped her into night-clothes and then into a narrow bed. Marta was asleep in an instant.

Despite the quiet night she slept fitfully. Her face was a mask of pain, and her fever waxed as full as the equatorial moon. The next morning Abuela prepared a simple meal, cornmeal cereal, fruit, and coffee. Marta picked at her food and stared at the bowl.

The small, wizened figure stood still save a crooning voice. "*Heee-jaaa,*" the old woman intoned, stretching out the vowels: child. A single word carrying six decades of love and wisdom.

"*Hija…mira.*" Look.

"At what?"

Abuela touched her hand to Marta's heart.

"At my shirt? At the button?" The girl's voice cracked. She regretted the sarcasm.

"*Estás tan airada.*" You are so angry.

"I'm just tired."

Abuela pointed to Marta's clenched fist.

"Why didn't you come, Abuela? Mom needed you. Now Dad is, I don't know, lost. He doesn't think straight."

Abuela said nothing. She took Marta's hand and began, gently, to uncurl her fingers.

"There's nothing you can do! Look at me. I have JRA. Do you even know what that is? It's juvenile rheumatoid arthritis and it's not going to go away." She sobbed into her grandmother's bosom. "What am I doing here? I miss her so much."

Abuela reached into the same leather pouch that she had worn at Rafael and Elena's wedding. It was the size of her fist and as worn as her sun-wrinkled brown skin. She took a handful of herbs she had picked at dawn—bright green leaves and deep lavender flowers—and placed them into boiling water as the girl's tears spilled. A savory fragrance enveloped Marta.

"Drink this," she commanded gently. Marta drank the liquid with a grimace. Warmth soon suffused her legs and they seemed to unlock, as if from their own volition.

Marta felt herself relax. "What was that?"

"The plant is called by many names. Here we call it *ajos sacha* or false garlic. It helps with swellings. The pain will leave you for a little while and you can walk with me in the forest to meet Yocahu. Hija, come with me," Abuela beckoned.

"Who's Yocahu?"

"You will meet him. El Yunque is his home. It is named for him."

Marta felt lighter. Her legs lost their unsteady gait and she moved more easily. The two women entered the rainforest. A lush green canopy stilled the wind, and the sea's gentle lapping was a distant obbligato, rhythmic counterpoint to the caws and twitters of the forest's exotic birds. The ground was covered by soft mulch, centuries of decayed leaves that muted their footsteps. Golden sunlight refracted through the trees overhead, bursting here and there into a rainbow of colors. Angle lizards skittered across the ground and up the trees. Marta could taste salt in the moist sea air and her skin cooled as the fever abated. She drew in an easy breath and was no longer hunched in pain.

"What's happening to me?"

"The pain leaves you and you are free to know yourself," Abuela said. "It is a gift from Yocahu."

"Who is this Yocahu?" Marta repeated. She felt lightheaded.

"He is the god of the forest, El Yunque. His healing plants are here."

"Will this cure me?"

Abuela walked in silence before she spoke. "I do not know. I will tell you the names of Yocahu's plants. I will tell you their stories. The rest you must find out for yourself."

"Are there other gods here?"

"There are many, including the enemy, Juricán. He is the god of the hurricane. He is an angry god and even strikes those who walk with him."

"I haven't felt like this in a long time. I wish my doctor knew about your medicine."

Abuela laughed, a pleasant sound, curiously basso. "It is not my medicine. But you are right. Hospital doctors do not know much about Yocahu's plants. You could teach them."

"Me?" objected Marta. "I'm just a kid."

"Yes, but what becomes of children? They become adults. What becomes of adults? Do they follow their hearts or are they filled with discontent? Why not do what's in your heart?"

"That's a kid's question?"

"Hija, it is the most important question. It is one that adults lack the courage to ask. Yes, this is very much a child's question."

They walked amid the plants and insects. Abuela touched Marta's arm. "Be careful not to step on *bibajagua*, the ant. He is a friend to the forest, but he can bite you."

Now Marta laughed, thin and reedy. "I'm not worried about an ant."

"Why not?"

"Look at it. Why would I care about something so small?"

"What about you? You are small. Your legs give you pain. Why would anyone care about something as small as you?"

"I'm a person, not an ant," said Marta.

"Is there a difference?" asked Abuela.

They walked further. From time to time, Abuela pointed to a flower or a shrub and explained how she used the plants' healing parts, the bark or leaves or roots or petals.

"How do you know all this?"

"I am a bohique. A medicine woman. I am Taíno," the old woman said.

"I thought there were no more Taíno people. Didn't Columbus wipe them out?"

"Perhaps we are another part of the forest's secrets. Columbus was the first Spaniard to find our island but he did not stay here. It was Ponce de Leon who enslaved us and caused so many deaths. He traded disease for gold. When we rose in protest he slaughtered us."

"Did he kill the Taíno?" Marta asked.

"Not all."

"But if there are still Taíno, why do the books say that they're all gone?"

"If there are no survivors, then there is no one to demand justice. So the records say we are no more. The records do not mention the places of the Taíno, like Orocovis, Caguas, or Yauco. Even in New Jersey and Florida you can find Taíno. The scientists say that there is Taíno blood in the Puerto Rican people, but they do not admit that the Taíno still live."

"How come nobody knows about this?" Marta asked.

"I know. Now you do, too."

They walked farther into the forest, past waterfalls and flowers, trailing coral reefs, beaches, lagoons, and mangroves. Marta heard the gentle cry of birds and the song of the tiny *coquí* frogs that had greeted her the night before. Abuela paused and Marta considered the old woman. A faint smile crossed her wrinkled face, and Marta saw something profound in her eyes.

"Listen well," Abuela explained. "Remember bibajagua, the ant. If you learn how to care about him, you will learn how to care for yourself."

"But that's just an ant," Marta said. "And he's such a little thing,"

"He's a little thing but precious."

"I don't know much about ants. They seem to do okay without me taking care of them."

"A bohique does not have to care *for* everything but *about* everything. If you are going to care about the little precious things, then you must even care about the little deadly things."

"What? Like, snakes? If I see a snake, we'll find out how good your medicine is. I would run so fast!" Marta laughed again.

Abuela smiled with her. "How will you know the difference between the precious things and the deadly things?"

"I know what a snake looks like."

"Can you see the snake's heart?" Abuela asked. "What about bibijagua? Is he precious or deadly? His bite is painful. A colony can strip the leaves from an entire crop overnight. But bibijagua brings fallen leaves underground and makes the soil rich. So if we try to stop bibijagua because of his bite, then we lose the life in the soil. You must be able to see the whole of bibijagua to know if he is a precious thing or a deadly thing. It is the same with people."

Abuela stopped and faced her granddaughter. She put her hands on Marta's shoulders and held her firmly. "The precious things and the deadly things grow together in this world. They grow together inside people. Can you destroy one without destroying the other? Look at yourself. Your mother was so sweet, but helpless. Your father is so strong, but confused. But could you exist without both of them? Your father's strength will give you courage to cope with your pain, and your mother's blood will help you be a healer.

Marta walked and considered her grandmother's words. Would she be crushed by her pain and her loneliness? She had no real friends. Other children ran and played but she was slow and lumbered. Her mother had the patience to walk with her, but her mother was gone. Maybe she was just an ant. But the tiny ant was powerful.

As Marta considered bibijagua, she felt a small change inside herself. It was as if the tumblers of a lock within her began to move. Their movement was fractional, but the distance they travelled was unimportant. It was their alignment that would let her grow despite her disability. Without realizing it, Marta determined that she would survive. She would find a way to thrive.

The old woman continued. "Here is the almacigo tree. You can use it to cure a stomach ache or diarrhea. And see this one that grows right next to it? This is the tartago, also for your stomach. Here is the cojobana tree. Her seeds will give you a vision to see the future."

They passed through the forest, returning as the shadows

lengthened. While her grandmother prepared supper, Marta fell into a pain-free sleep, her first deep slumber in months.

The hours and days and weeks passed for Marta. The soles of her feet toughened and so did her will to survive. Her skin had bronzed and radiated health. Marta was an apt pupil, hungry for knowledge. She learned the names of the flowers and trees and birds and animals and insects. She learned what leaves might cure a headache or fever and which ones might still a child's crying. She watched intently as Abuela prepared breakfast and dinner. Soon, Marta prepared the meals, then the medicines. She was becoming a bohique.

As the pain in her legs diminished, she stood straighter and walked with confidence despite her awkward gait. Her neck looked too slender to support her head but Marta held her upper body erect, perhaps to compensate for her limp. She tossed her raven hair and it danced on her shoulders. Soon she would brush it to a luster and experiment with style, but for now she treated it with a child's abandon.

As the pigment in her skin deepened under the equatorial sun, her teeth shone all the whiter. A beautiful smile escaped the custody of a once-perpetual frown. Marta radiated delight and unrestrained curiosity as she learned the lore of her people.

The days shortened just a little as the tropical sun moved to what passes for autumn so close to the equator. The temperature remained constant. Most of the season's changes were subtle but Marta saw them. As the summer waned, Marta's stay with Abuela drew to a close.

"Abuela, I don't want to leave."

"But children must go to school," Abuela said. "And you will have a special school."

Marta had been chosen to attend a charter school in East Los Angeles. Students from around the world would join Marta in an experimental school program created by the Hidden Scholar Foundation. It sought children who had two things in common: poverty and brilliance.

"When will I see you again, Abuela?" the girl said, choking back tears as she packed for the long flight home.

"You may come and visit anytime but you can see me whenever you look at the trees or the sky. I walk with Yocahu and so do you," the old woman said.

Marta embraced her grandmother and hugged her fiercely. "I love you so much, Abuela. Thank you for teaching me."

"You're welcome, child. Remember what you learned. Remember bibijagua, the ant."

"I promise I will. Abuela, before I have to go, please tell me one thing. When you married Mom and Dad, they had a vision. What did it mean?"

Abuela took Marta's hands in her own. The carro publico was waiting to take her to the airport but Marta would be the first passenger and the driver was in no hurry. Marta could still hear the coquí frogs and the rhythmic whisper of the Caribbean's small waves.

"Juricán will touch you," the old woman said. "I do not know how. This is the meaning of the golden vine with the black strand. Juricán will come, not as a spirit, but in flesh and blood. You will have your own protector with his own knowledge and he will be tempted by Juricán. He may follow the hurricane or he may not. And a golden strand will grow from you as well, one that will know both Yocahu and Juricán.

"But you must take the knowledge you found here to the doctors of your world. These plants will disappear and the knowledge of the bohique will be lost. You must bring Yocahu's gifts to the doctors of your world."

Marta thought about Abuela's words to her. A child teaching scientists about Yocahu? A battle with Juricán? It seemed farfetched.

"Abuela, the doctors aren't going to listen to me when I tell them about plants. And how am I going to fight a god?"

"Hija, you know almost as much as any bohique. Yocahu has given you this knowledge and you learned it well. My heart sings to watch you grow."

"I can't say that your prophesy makes me feel very optimistic," said Marta. The sarcasm that ebbed over the summer crept back into her words. "Let's see, I've got disease from my mother, a helpless father, and a battle with the God of Evil in my future. Is that it, Abuela?"

"No, hija. There is one other thing," the old woman said.

"Oh, great," Marta muttered and rolled her dark eyes.

Abuela smiled. She reached behind her neck and her fingers worked for a moment to untie a knot in a leather cord. It was attached to a leather pouch she carried between her breasts, next to her own medicine bag. This one was older, tanned more deeply. A delicate image was burnt into the leather, a branch with twenty-four long, thin leaves. Marta recognized the leaves of the cojobana tree, giver of visions.

"I was saving this for the right moment. I think that is now." The old woman grinned.

"This was my mother's. Now it is yours. This is part of your legacy, too. Pain and healing dwell within you. Give each one its voice, but do not let one drown out the other. And do not let these voices drown out your own voice." The old woman's arms encircled Marta and hung the pouch around her neck.

Marta hugged her grandmother and breathed in deeply. She

closed her eyes and fixed the image of the twinned vines of her parents' legacy. Her meditation shifted to the golden strand of her own life, and of the one to come. She visualized growth, impervious to the black filament. Her vision expanded to include the rich soil of El Yunque nurturing the roots of her vine. She felt powerful, connected. The spirit of the forest was substantiated within her. Then she walked away to the publico with grace and purpose and turned back once more.

Abuela called out, "Hija. Your mother was always proud of you."

Then the old woman vanished into the forest.

04

A BOY AND HIS DOG

Jim Ecco, age thirteen, and Ringer, age three. A boy and his dog. On the good days, Jim and Ringer visited the Pasadena library. Ringer waited at the entrance and ignored slinking cats, curious dogs, nervous passersby, restaurant aromas, and branch-borne squirrels, although that was difficult even for a Good Dog. On the better days, Jim could slip into the passages of the books he brought home on his dataslate or on paper, and his own world disappeared. On the bad days, Jim and Ringer curled up together and listened for the weight of Dad's approaching footsteps.

Ringer was a mutt, Heinz 57, as far from the show ring as a stevedore from a fashion runway. She was part terrier, brave and independent, part German Shepherd, protective. Her coarse undercoat resisted brushing and shed uncaring torrents of light brown hair. Jim's mother vacuumed it from the sofa and the carpets. She cleaned hair in the kitchen and the family room and especially from Dad's chair.

Jim was thirteen but displayed few marks of maturation. Classmates teased him for his soft looks. His Adam's apple, cheekbones, and jaw line were still undefined, and framed by a mop of sandy brown hair, as unresponsive in its way as Ringer's.

Mom called Jim. "Clean up before your father gets home. Let's have a nice dinner tonight."

Mom's voice sounded strained and Jim guessed Dad had called ahead, in a mood. Galvin Ecco, his father, was an attorney. Lawyers in the movies were smart and always in control. Lately, Dad was not in control. He stared and snapped, and then there would be a reckoning. Sometimes just a slap, sometimes more.

When Dad talked about law, he was at ease. "A thing is, or it is not," was his favorite saying. Mom said he was too rigid and that was bad for business. Jim didn't know who was right but that more and more, Dad lost his cases, his clients, and his temper.

"Jimmy. Clean up now before your father gets home."

Too late. Dad arrived. He walked as stiff as a man with a rash and wore a dark navy suit with faint white pinstripes and frayed cuffs. His hair was shaped into a precise crew cut that would please a drill sergeant. Its color reminded Jim of a thunder cloud.

Jim watched as Dad looked left and right, peering over his glasses. "I can't control the courts," he often said, "but the damned house better be clean."

Dad's first words to Jim were, "Is your room neat?"

"Neat as a pin."

"I'll be the judge of that. I expect your room to be immaculate. You know what immaculate means?"

Jim said nothing. Housekeeping was a herculean task for the thirteen-year-old dreamer, and his father exacted a military standard.

Pinstriped dad and dungareed lad marched to the child's room. Jim's bed was in the far corner. Wall-mounted bookshelves crowded a desk and straight-backed chair. A nightstand supported a reading light and Jim's current printed fare, a pile of old graphic novels, tales of amazing feats and dark retribution. Ringer lay on the Berber carpet. Mom had said that dog hair would be less noticeable on the tan and grey weave, but the carpet's geometric pattern seemed to showcase every bit of dirt or scrap of paper, every piece of furniture even a degree off-square.

Two years ago Jim enjoyed a larger room in a larger house. Then the family moved into smaller quarters. When Jim asked why they were moving for the second time in as many years, Mom smiled and said, "We're saving for the future." She took too long to answer.

When Dad walked in the room, Ringer's shoulder muscles bunched, her weight shifted to her hindquarters and her ears pulled forward. Now she looked more like a shepherd, protective, curling half into a prey bow, rather than her happy-go-lucky play bow.

Jim fidgeted while Dad inspected. The books were arranged on the shelves, from tallest to smallest: Dad's Rule. School supplies in a pencil cup, pens down and pencils up. Dad's Rule again. Clothing was put away. No litter on the carpet.

Then Dad looked under the bed and found a tangled clump of dog hair. Jim didn't think anyone else in the world would care but Dad acted as if it were a malignant mass poised to metastasize, to cover Jim's room, the whole damned house, with canine detritus.

Some other day Dad would understand that under the bed doesn't count. He would sigh, shake his head, and dismiss the furry tangle. Or Mom might intercede, "Galvin, the boy has homework.

Let me finish so he can get to his studies." Dad might let it go. Or maybe not, and Jim might hear them argue—or more. They were like dancers in a tango of insults and hands. He could picture Mom. Her words were her weapons. She leaned into Dad's blows, to store each impact and then return his fury with her own taunts and barbs.

Dad held the offensive find between the tips of his thumb and forefinger. He glared stony-faced at Jim. Ringer's ears pricked.

"You too lazy to vacuum?"

Dad held the mass, extended his hand, and dropped it back on the carpet. "You think this is clean?" He swept his right arm across Jim's desk, knocking pens and pencils to the floor.

"I'm sorry. I'll do it again."

Too late. Dad's face reddened and he picked up his son, his flesh and blood, the vessel of his hopes and dreams. With one arm around the boy's chest, the other around his legs, Dad held him head first like a SWAT officer might hold a battering ram, poised at a felon's front door. He swept the boy across the desk. Jim's books, tallest to smallest, scattered.

"You damned well better learn to clean up after that goddamn dog."

It's not Ringer's fault, Jim thought, but kept silent—no way to know how Dad might react.

That was the problem. Jim never knew what to expect from his father. He thought of last summer's family vacation. They drove the rocky central California coast through Big Sur, north toward San Francisco. The narrow ribbon of road hugged steep cliffs and presented spectacular ocean views. Jim peered down to the Pacific and back to the car's odometer, counting down the miles to Monterey. The Monterey Bay Aquarium drew him as surely as a siren's call. Never mind that the Parkfield earthquake destroyed half of the collection just nine months earlier. The Kelp Forest survived and Jim was eager to see the thirty-foot fronds sway in an oversized tank.

What set off Dad that morning? Maybe it was the traffic or something between his parents. They seemed to have a special language—one with unspoken shades of anxious meaning, an emotional carrier wave under plain words. As Dad instructed his car to pay for parking, Jim urged his father to hurry. There was a whole forest of kelp to see. Dad turned and slapped him. Not too hard, nothing that would leave a mark. Dad called that his Simmer Down Slap.

From up the street, someone yelled out, "Hey! Leave the kid alone!"

Dad ignored it, but not Jim. This was a family affair. Before he could stop himself, Jim yelled back, "He can hit me anytime he wants!"

Uh-oh. There's going to be heck to pay for that one. But Dad's shoulders drooped. "Never mind," he said. "Let's go see the kelp. Just don't talk like that again, okay?"

Dad was quiet that day, even kind. But that was Dad. He might beat Jim with a belt, and often did, but then he was quick-witted, engaging, eager to explain how the world worked.

But not today. Not when Jim failed inspection.

As he left Jim's room, Dad aimed a kick at the Ringer's hind-quarters. The dog scampered out of reach.

"Please don't hurt Ringer. It's not her fault." Frustration and rage were boiling inside of him and he struggled to control his voice.

"Don't *you* tell *me* what do!" He turned and stepped back toward the dog.

Too much. Jim took three fast steps to stand in front of his father and screamed, "DON'T YOU TOUCH MY DOG!"

Dad looked startled. "Or what? How are *you* gonna stop me?"

Jim's hands shook but he clenched them into fists. Dad raised one hand in a warning but Jim stood his ground. After all, there is something about a boy and his dog.

"Get to work. Clean your goddam room."

The door slammed behind Dad and Jim knelt to hug Ringer.

"It's okay, girl. It's okay." He shook as the adrenaline in his bloodstream tried to activate every muscle in his body.

Jim caressed the dog's long, smooth muscles, running his hands from her withers to her hips. Ringer's ears moved back as the effleurage calmed them both. The simple act of stroking the dog—and being stroked by the boy—triggered a release of oxytocin in both the boy and his dog. The hormone enhanced their bond and calmed them.

Jim stroked Ringer from shoulders to brisket, collecting dog hair as he went. He twisted it into a ball so that it would not litter his room, adding it to the offensive mass discovered under the bed. Later that evening Jim pushed the clump up into the muffler of Dad's car. The next time his father drove the car long enough for the metal to heat, the dog's hair would smoke. No harm to the car, no evidence of Jim's payback—save the stench of burning hair, brief enough to be inexplicable, strong enough to make his father gag.

At bedtime, Ringer curled at the foot of Jim's bed. Mornings, Jim woke to find Ringer's muzzle perched inches from his face. Did she stand sentry all night? How else would Jim awaken every morning to the sight of two soft, brown eyes?

One morning, he awoke slowly, wrapped in the helpless pleasure of sleep's immobility. He imagined that he was an Indian papoose, swaddled and strapped to a cradleboard. Safe. Ringer was still at the foot of the bed. Jim's breathing changed as he emerged into wakefulness. Ringer stood, stretched, front legs down and hindquarters up, her back bowed. She took her customary post, snout resting lightly on Jim's bed.

She hears my breathing change. She hears me wake up. That's how she does it. What else does she notice? he wondered. *I'll watch her and learn.*

He learned to react with Ringer. She alerted him to the subtle signals of Dad's anger, like the tightening of his neck muscles. When Dad was in a mood, Ringer's ears snapped erect. Then Jim saw his father's skin flush with anger as clearly as a lighthouse beacon. He saw the flare of nostrils, the widening of his pupils, the shift in balance. Dad had a tell, like a poker player staring too long at a hole card. If Dad rubbed the back of his neck when he was angry, then he was about to lose his temper.

Ringer reacted to Mom, too. Why? Mom never yelled or hit. She might scold Dad—mostly about money—but she never lost her temper. But Ringer's ears pitched forward anyway and now Jim noticed the tension in her smile.

Sometimes she provoked Dad. Her words weren't so bad and she never used swears. But Ringer reacted and Jim listened. He heard acid-laced tones, derision in Mom's voice. When she combined a certain intonation with a particular cant to her body, Dad would react, hands flying. It was as if he had a mad switch and she closed the circuit. Then Dad struck.

Jim learned to move like his dog. Ringer's head was like an arm whipping this way or that to deliver a canine mouth at play or prey. Jim's arms learned to deliver his hands as well. Ringer's mouth was both delicate and powerful. He could carry a baby bird, fallen from its nest, or grind a marrow bone to a sliver. Jim's hands learned tenderness and anger. The boy who had discovered every plane, curve, and hollow of Ringer's form began to learn the strengths of his own form and the weak spots of others.

Now Jim could dodge Dad's slaps and blows. But a slight, thirteen-year-old boy is no match for an adult. Jim was fast, but he would tire, and Dad never got smaller. The odds favored size, and the day before Easter vacation, Jim's luck ran out. He was cornered in his room.

"Where you gonna go now, little man?"

Jim checked Dad's hands. They were open and empty. Ringer was not in the room. He faced Dad alone.

"I asked you a question. Where you gonna go now?" Dad lunged and Jim ducked under his father's arms.

"Have it your way. But remember this is my damned house." Dad's mouth curled into a smile and then left the room. Later, Jim would remember that the smile never reached his eyes.

Jim whistled for Ringer and they slipped into the Pasadena evening. When they returned, Jim opened his bedroom door to a near-empty space. His books were gone. His reading lamp was gone. There was a cot in place of his own bed. Even Ringer's bed was gone. His father stood in the doorway.

"You think you're smart. Well, remember that this is my house and I pay for everything."

Jim's last thought before the tears fell was, *Well, I guess I can go to the library.* He felt helpless, powerless, diminished by his father's insult, "Little Man." *Something's got to give,* he thought, *or I'm going to go crazy.*

When he regained his composure he walked into the kitchen, opened a drawer, and removed a box of toothpicks. He placed two of the wooden slivers in his breast pocket and slipped into the night, blind to the world around him, operating on habit alone.

Had Jim looked up, he could have traced the forms of the

constellations. It was early evening and he might have looked for Libra. But eyes were cloaked in anger and his vision fixed into a narrow spot on the sidewalk just in front of him. On another evening, he would have delighted in the scents of Southern California's abundant flora, but tonight, even the night-blooming jasmine smelled cloying. He heard neither the whisper of the evening breeze, nor any sound except blood pounding in his ears.

Dad's workplace was two miles away, thirty minutes at a schoolboy's angry pace. He approached the front door of the storefront office and removed one of the toothpicks, broke it in half, and inserted a piece into the tumbler mechanism of the door's lock. He used the other toothpick to push the broken piece in as far as possible. Tomorrow Dad would be locked out of his office and the entire lock would have to be removed and replaced.

A security camera recorded every move.

The next day, the last school day before Easter vacation, a pulse still twitched in Jim's neck. He ignored greetings from teachers and students. He ignored the bells that signaled the change in classes, navigating by rote. He ignored his lunch and moved to his afternoon classes with all the focus of a man in a coma.

The trance broke during math class. The teacher was administering a quiz. Jim sat unmoving.

"Mr. Ecco, would you like to join the rest of us in the exercise?" She smiled.

Jim did not reply.

"Mr. Ecco? Jim? Are you all right?" Her voice was bright, but with a note of concern.

The teacher walked down the aisle to Jim's desk. When she reached out to touch the boy's shoulder, he saw his father's hand. He heard his father's voice. Jim's arm flew up and knocked aside the teacher's hand. In the same motion, Jim stood, too quickly, and his desk tumbled over. The edge scraped down the woman's shin. It was painful but not damaging. Still, it would cost Jim the rest of the school year.

Jim looked at his teacher. "I'm sorry," he said, and left the classroom. He walked home, into his bare room, ignored the cot and lay down on the floor with Ringer, unmoving, until the police arrived.

On the following Tuesday, school principal Danny Sorenson sat in a tan club chair that was browned from use, the man's form outlined in darkened leather. Sorenson was in that indeterminate middle age when his belly had begun a winning battle with his hair for prominence. He wore a red bow tie, a white shirt, and a forest-green cardigan sweater vest and rumpled khaki pants.

Jim sat on a matching sofa, opposite the administrator. He'd been there before. Sorenson had asked about Jim's home life, had reached out to Jim and tried to find some activity that would help Jim channel his frustrations. "You're a smart kid," Sorenson said. "Your aptitude tests say you've got a lot of potential."

But today the conversation would be about survival, not potential.

"Jim, you're in a pickle," Sorenson said, not unkindly.

"I'm sorry," said Jim.

"The incident with Ms. Rice was reported. She says that it was an accident that the desk struck her leg, but when you hit her arm, technically, you assaulted her. Can you tell me why you did that?"

"I don't know."

"The police are considering dropping the charges against you."

"Whatever."

"No, not *whatever*. Jim, this is serious. Your father is waiting outside. He needs to be part of this conversation but I wanted to talk to you first. Jim, what's going on at home?"

Jim said nothing.

"Okay," Sorenson shrugged. "Let's get your father."

When Sorenson brought Galvin Ecco into the office, the attorney glared at the principal, glared at his son, looked around the office and, for good measure, glared at Sorenson's framed credentials.

"Mr. Ecco, you're an attorney. Can you explain to your son how serious this is?"

"No."

"No?"

"It's his mess. Let him fix it. Are we done here?" Galvin rose to leave.

"No, Mr. Ecco, we are not done here. Please sit down. There's a second problem, one that involves you directly."

"I don't like the tone of your voice," Galvin said.

"Sir, I'm sorry you're upset. But your son is going to be expelled. It's school policy."

"That's his problem. He also vandalized my office. Did he tell you that?"

"It sounds like he's pretty angry about something. Do you know what that might be?" Sorenson asked.

"I don't know and I don't care."

"Mr. Ecco, the question is, what are you and Mrs. Ecco going to do about Jim's education? If we can show a plan for rehabilitation that includes keeping him in school, the police will drop the charges. But he's not going to be able to return to this school."

"So, what's going to happen?" Jim asked.

"Well. That's why we're here," Sorensen said.

Jim's father raised his voice, "He vandalized my office, he hit the teacher. He's a big boy, he can pay the price. He's got to learn some discipline."

"Mr. Ecco, can you do me a favor?"

"What?"

"Settle down for a few minutes? Every family has problems. But yours cross over into my school and you can't just wash your hands of the matter. Your son is thirteen years old, and you're responsible for him."

"What the hell am I supposed to do? He crossed the line with this stunt."

"I'm trying to help, Mr. Ecco," Sorenson said quietly. Then, a bit sterner, "Now please listen." Galvin's face colored. He opened his mouth and closed it, then opened and closed it again. For the first time since his books were stripped from his room, Jim became animated. A half-smile turned up one corner of Jim's mouth.

Sorenson looked at Jim's father. "Here's my proposition. I've arranged a transfer to another school district where your son can start fresh."

"Where?"

"Los Pobladores High in East Los Angeles."

"East L.A.? Some ghetto school? Let's see how smart he can be down there."

"Actually, Mr. Ecco, Los Pobladores would be a good school for Jim. It's one of the schools sponsored by the Hidden Scholar Foundation."

"What's that?" asked Jim.

"The Foundation takes good students from poor neighborhoods around the world. It places them in low-income neighborhood schools in the U.S. and then provides funding to those schools. The Hidden Scholar Foundation is the creation of the philanthropist, Robert Murray Herbertson."

"The rich guy?" Jim asked.

"Yes, the rich guy." Sorenson stroked his chin and his eyes went back and forth between the father and son. Then he fixed his gaze on Galvin. "Mr. Ecco, your son won't be a Foundation scholar, but he will benefit from the Foundation's programs. I've arranged for him to transfer to Los Pobladores. I know the principal there and we worked out an arrangement. We do this from time to time when a change of location might benefit a good student."

"Jim is *not* a good student," said Galvin.

"He's an underachiever, but he has a lot of potential."

"Well, I'm not driving him all the way down to East L.A. every day. And there's no train from Pasadena to East Los Angeles."

"Actually, sir, in view of the, uh, tension, at home, we've arranged for him to board with a local family—with your permission."

"What about my dog?" said Jim. "What about Ringer?"

Sorenson sighed. "You're going to have to work that out. Right now I'm trying to keep you out of the court system." Sorenson unrolled his dataslate. Jim saw his school records. Sorenson continued, "Jim, I think you can make something of yourself, but you have an attitude problem. In the last nine months, you've been in three fights with other students."

"It wasn't my fault! I never start it."

"I know, but each time you could have walked away."

Jim started to protest but Sorenson held up a hand. "Stop. You have an attitude problem that's getting you in trouble. Part of the plan to clear your record involves that you be placed in another home for the school year, if your father consents. Let's see if that makes a difference."

"Mr. Ecco, if we take this action, the courts will be satisfied. Your son will not end up with a juvenile record, and you avoid liability if the teacher seeks damages. As an attorney, I'm sure you can see the benefit to you."

Turning back to Jim, Sorensen said. "Son, no matter what your father decides, you're out for the rest of the year. You're going to have to attend summer school to make up the days you miss here." Jim heard a tone of finality in the principal's voice.

"That's not fair," Jim protested.

"Enough! You assaulted a teacher. I know it was an accident, and you didn't hurt anybody. But it was reported to the police, and this is the way it's going to be."

"Who reported it?" asked Jim.

"What difference does that make? There was a class full of students, and students talk. Ms. Rice needed some treatment for the scrape on her leg. So, there's the infirmary. Someone might have been walking by. It doesn't matter now. Keeping you out of the court system is the most important thing. Mr. Ecco, will you allow Jim to board with another family so he can attend Los Pobladores? If you agree, Jim's record gets expunged and you won't have to worry about a lawsuit."

Dad said, "Yes. Are we done now?"

"Yes, Mr. Ecco, you and I are done." Sorenson sighed again. It had been a long day, a long weekend, one that started when he picked up the phone, called the juvenile authorities, and arranged for Jim's arrest and for his reassignment to a different school and a calmer home.

❁   ❁   ❁

Jim completed summer school at Los Pobladores. In the fall, on the first day of classes, his attention was drawn to another freshman student, otherworldly and beautiful. She spoke with a Puerto Rican accent and walked with a limp.

05

SCHOOL DAYS

At 7:30 AM, Tuesday, September 6, 2022, two students emerged from the Hidden Scholar Foundation car. The girl with the black hair craned her neck to take in the neighborhood, and then walked slowly towards the schoolyard. The smaller girl walked with an expression that showed simultaneous determination and disinterest.

Across the street, a trio of older students slouched outside a diner. They watched the two girls with undisguised hostility, appearing to agree on a course of action with raised eyebrows and nods. Their postures radiated contempt, and patrons emerging from the diner gave them a wide berth.

They were bullies with a grudge. Any of the eleven hundred or so returning students knew to avoid the stocky, pock-faced leader of the three who called himself Padron, 'Boss', as well as his cohorts, Frank Chung and Jamie Ortiz. Their prey were students from the Hidden Scholar Foundation. Targets of opportunity.

Padron had dull obsidian stones for eyes, broad cheekbones, and generous lips set in a puffy frown. Chung was stocky, his head and eyebrows shaven. He had a pictogram tattooed on his left cheek, a triangle containing a lightning bolt, the symbol for high voltage. Ortiz was tattooed with a rosy profusion of adolescent acne. All three affected a retro look, wearing long-sleeved shirts buttoned at the collar and low-slung baggy pants.

Padron acted first. He broke off from his two friends, crossed the street, and jogged around the building to confront the girls as they walked towards the school yard. Chung and Ortiz peeled themselves off the diner's wall and followed the two girls. They paced themselves to catch the girls in a pincer. Padron would approach from the front, they would close in behind.

A slight, tousle-haired boy stood outside of the diner with a companion. They also shared an easy familiarity and communicated with looks and gestures. The boy's gaze lingered on the girl walking with pride and a limp. He walked towards them. The boy's companion remained, comfortable where she was.

A mural that covered the length of the building drew the girls' attention. It was a panorama that depicted the community's history, starting with the arrival of forty-four *pobladores,* the original settlers of Los Angeles. It was a history of the city, of the neighborhood, and celebrated the birth of the high school. It was natural to walk past the colorful wall. Natural, but dangerous. That side of the building was windowless, perfect for muralists—and predators.

"Let's go look," said Marta Cruz as she turned towards the mural.

Eva Rozen dismissed it. "Who cares? Is painting. I want science lab." Her speech carried a guttural cadence that marked her Slavic pedigree.

"Well, I want to see it. You have all year to see the lab."

"You have all year for pictures."

"Yes, but classes don't start for a few minutes yet and we can look at the mural now. Let's go." Eva shrugged. She followed Marta to the painted side of the building. They were unaware of the eyes that followed their slow progress.

The fresco depicted a row of men and women dressed in rough-textured shirts and flowing robes, each settler pressed against the next. The figure in front held up a scroll with the words, "*Debemos ser libre*"—We must be free. At the top of the mural a large bird floated above a bronze-skinned man. He had the angular features of the area's indigenous people. Other figures carried guitars and accordions, scientific devices and crops.

Eva gave the mural a cursory inspection. The Pollock and the Dalí prints in Coombs's shop were more interesting. True, this art was more literal, but she returned to the works the antiquarian displayed in his office. Those were abstract, but somehow very personal.

Eva hung back and so was first to sense Chung and Ortiz behind them even as she saw Padron approach. She looked at Padron. Now her expression was equal parts disinterest and contempt. Marta Cruz's face showed open curiosity.

Patron appeared momentarily taken, perhaps disappointed by the girls' lack of fear. Then he said, "*Mira*." Look. "A cripple and a geek."

Chung and Ortiz took up station from behind, completing the pincer movement. Eva Rozen reached under her shawl and took out and shook a small squeeze bottle. Marta Cruz said, "What are you doing, Eva? Don't make trouble."

"*Oyé chica*, you got no trouble," said Padron, sliding forward. "Just show some respect, eh? Time to pay up."

Eva's gaze fastened hard on Padron and then shifted to the others. She made a mental calculation of the distances and shook her head at the disappointing conclusion. As she took in the unfolding scene, she noticed the slight figure of a boy walking towards them. He had soft, unassuming looks and seemed to draw into himself as he walked towards the confrontation. He looked too young for high school. Eva wondered if he belonged in middle school. "Hey," he called out to the girls as he approached. "Class is about to start. We've gotta go." He appeared oblivious to the trio's menace. To Padron, "How are ya, amigo?"

"I ain't your amigo. You're in the wrong place, *amigo*. You gonna pay some respect then you gonna get outta here. Empty your pockets, *amigo*."

"My pockets? Which one first?"

Padron looked hard at the interloper. "You funny?"

Eva calculated that Padron outweighed the newcomer by forty pounds and stood six inches taller. The boy drifted a few steps to the right, to Padron's left, his weak side.

Eva saw what Padron missed. She looked at the thin boy. "I don't need you to help," she said. He'd need a miracle to do what she believed he was planning. Then, to Padron, "Go to be someplace else," she said. Her accent and syntax helped her to sound bored.

Padron laughed and motioned to his friends. "Hey, Chung, Ortiz," he said, "We got us a party." They closed ranks.

Padron eyed the boy. "You a hero? That it, man? You gonna be a hero with no teeth."

The boy inched towards Frankie and Ortiz. Padron followed. The three older boys were drawn into a tight bunch. The hero looked off to his left, into the distance. Eva followed his gaze, but saw nothing. The hero's companion was hidden in shadows cast by the low angle of the morning sun.

Padron spoke again. "You gonna spit teeth, you don't turn your pockets out now."

The hero smiled. Likely, the smile was intended to look disarming rather than demeaning. The smile hid the years of accumulated frustration and rage. Behind the soft face, he boiled. His smile broadened.

Padron snorted. "What's so funny? I don't think you're funny."

The hero turned to the two girls and said, "I think you better get out of here. Maybe you should run into the school."

Marta Cruz wore an expression of amusement and contempt. She gestured to her legs. "I'm supposed to run?" She rolled her eyes, shook her head, and spat out the next words. "Boys. All the same. All brave and no brain."

Eva said nothing, her eyes still calculating distances.

Suddenly Padron wound back like a pitcher on the mound at Dodger Stadium. The hero watched calmly and ducked easily. "Get out of here," he shouted again at Marta and Eva.

At that moment, the boy's companion, watching from across the street, underwent a metamorphosis. Her ears pulled back and her lips drew forward. She dug her hindquarters into the ground, driving forward, front legs extending to double her length. Her body was low to the ground and she looked like a fur-covered missile, tipped with a toothy snarl. She hit maximum velocity in two strides and then covered the seventy-foot distance to Padron in less than three seconds.

She was in the grip of instinct and drive, a terrier's lust for prey and a shepherd's need to protect. Her tail was low to the ground for balance. Adrenaline flooded her, amplifying behaviors that had been hardwired into her species for millennia. Her lips drew further forward into an aggressive pucker. Sixty pounds of focused motion covered by a wiry tan coat. An unexpected white band circled her tail, the inspiration for the name to which she responded: Ringer.

Ringer's nostrils flared and closed rapidly, forcing scent molecules to receptors deep in her brain. There two enormous olfactory bulbs sorted the smells of the group and passed commands directly to her muscles. Her specialized scent organs freed the slower forebrain to calculate distance, velocity, and vector. The stink from the tallest of the targets, pheromones of fear and excitement from the girls, were as easy for her to read as a billboard would be for her two-legged companion.

Eva watched the dog. Six feet from Padron, Ringer's back legs drove her up, propelling her full weight into the chest of the surprised leader. Eva thought the dog was grinning.

Some of the blow was cushioned as the canine joints flexed. Still, the force was enough to knock Padron hard to the ground. He landed on his back with a whoomp. His diaphragm muscle spasmed on impact and prevented him from drawing air into his lungs. When he opened his eyes, his view of the world was circumscribed by a set of canine teeth inches from his face. By the time Padron could draw his next breath, the encounter would be over.

Eva saw a smooth blur of motion as the hero turned to the downed Padron. He tensed his body and aimed a powerful kick at Padron's ribcage.

Padrone rolled in pain. The Hero's foot missed.

The momentum of his failed leg strike pulled him off-balance and he landed on his back, next to Padron. The two lay staring at each other. They both gasped for air, fish thrashing in the bottom of an angler's boat.

Chung started to laugh. "Oh, man, this is too good," he

managed. Ortiz merely gave a half-grin and snorted. He nudged his friend and pointed to the prone hero. He said, "Hey, Chung, what we do about this *pendejo?*"

"Let Padron do him when the two of 'em are finished with their little siestas," Chung said. He turned back to Eva and Marta. "You still need to pay a tribute." He folded his arms and glared at the two girls.

"Okay," Eva said. "I got nice present for you." Eva said. She thrust the small plastic bottle she'd taken from her pocket moments earlier and squeezed a stream of oily liquid into Chung's eyes. It was a perfect opportunity, she had decided, to experiment with her new pepper spray. Could the effects of the local Habanero peppers compare with her treasured Guntar peppers? She observed that the heat from the southwest Indian peppers was more potent, but the Habaneros lived up to their reputation. They burned.

Science in action, Eva thought with a grin.

Chung yelped in pain. Padron was still on the ground, his view of the action limited to the forty-two canine teeth directly over his face. Ortiz had stopped laughing and looked puzzled. By now the hero was up, fists clenched, his face a twisted in rage. Eva wondered if he was going to have a heart attack. The boy tensed to strike Ortiz, but Marta Cruz stepped between them.

"Stop. There's no more fight. Let it go," she said. Eva wasn't sure whom Marta was addressing. The hero checked his motion and struggled to keep his balance. Ortiz stood, a bemused look on his face.

Marta turned and knelt at Padron's side. He was still struggling for air. She knelt and grasped the front of his waistband and lifted his hips sharply several times, until his diaphragm relaxed and he could once again draw air in panicked gulps. Marta turned to the hero, and asked, "Would you ask your dog to let him get up?" The hero gestured, his palm extended as if he were a bellhop waiting for a tip. He brought his hand up ninety degrees, his palm facing inward like a backwards hello. Immediately, Ringer sat.

Marta turned her attention to Chung. She found a bottle of water in her bag and drenched his eyes, then turned to Eva. "What was that?" Marta asked.

"Pepper spray. I make."

Marta handed Chung the water and told him to rinse but not rub his eyes.

"You didn't have to spray him," Marta said.

"Nobody attack me without hurt."

Padron stood up, wary eyes fixed on Ringer. Ringer drew back her upper lip. Padron backed up a step. He caught Ortiz's eye and nodded. They grabbed Chung and started to walk away. Padron turned to the hero, "You know what? I'd have kicked your ass except for that dog. Sometime, you and me? We gonna meet up again, no little girls to protect you."

The hero said, "If that's what you're going to do, then that's what you're going to do. But no one meant you any disrespect and if everybody stays cool, then nobody finds out that you got *your* asses kicked by a dog and a little girl."

"You crazy, man," said Padron, but the fight had left him and his threat lost its menace. He walked away.

Marta turned back to Eva, "You didn't have to spray that boy. You weren't attacked."

"Is technicality. He would attack but this one comes along." Eva turned to the hero. His face was soft again. "Your dog fight better than you. How you teach her that?"

"I didn't. She's never done that before."

Eva continued, "Where I come from, dogs is bad news. Dogs runs loose and kills."

The boy gave Eva an appraising stare. "Her name is Ringer. Don't worry about her."

"Dogs come to school in America?"

"No," he said. "She stays in the neighborhood during school, at least that's what we did during summer school. There are a couple of shops where they let her wait. I'll have to leave her home now."

Eva pondered. "She not bite. Why no bite? Is better with blood, yes?"

"Like I said, she never did that before."

"Whatever. That was good." Eva looked at the boy slowly, her gaze taking his measure. "I don't like dogs but this one, maybe okay. You helped us. I say thanks to you. I am Eva Rozen, this is Marta Cruz."

"Jim Ecco."

"What kind of dog is Ringer?" asked Marta.

Jim shrugged. "Some terrier, maybe. Possibly an Airedale, from her size. German Shepherd? Who knows?"

Eva approached Ringer, hand outstretched. "Nice doggie?"

Ringer backed up a pace.

"Hi doggie. I say, 'hello, doggie.'" Eva stepped forward again. Ringer backed up further.

"Dog is afraid of me?" asked Eva.

"Not exactly," explained Jim. "It's your posture. She doesn't like it when you lean over her with your hand stretched out like that. To a dog, that's rude and your hand over her head might be a threat. Just stand straight, relax and angle your body away a little. Like this." He demonstrated a neutral posture, "She'll relax. Don't face her directly until she knows you. And bring your hand up from underneath to scratch her chest."

Eva tried it and Ringer inched closer, sniffing. She allowed herself to be petted and then licked Eva's hand. For the first time since leaving Gergana's grave in Sofia, the Voices were silent and Eva Rozen smiled.

"Well, I'll be damned," she said, "What a nice dog." Eva's English was letter-perfect and unaccented, if a bit clipped, in the manner of one who had learned the language by rote, repeating phrases and vocabulary along with a recording.

The rest of the day passed without incident until the last period. The three students found themselves together for an English class. News of their morning confrontation had spread, despite Jim's assurance that it would be a secret, and classmates kept their distance out of deference or apprehension. Eva sat next to Jim, staring openly at him. Marta seemed focused on her classwork.

The English composition teacher was Henna Erickson. Her appearance was a nod to the styles of an earlier era—cotton peasant blouses instead of color-changing modern nanotextiles. She chose granny glasses to complete the look. Medium height, plain-faced, she had an unadorned figure draped in a shapeless dress. Her frizzy brown hair was pulled back at the nape of her neck.

In contrast to Erickson's utterly commonplace personal style, her classroom was animated. Action and emotion leapt from candid photographs on the walls: kinetic depictions of people at work, at rest, and at play. There were tender interchanges, confrontations, affection and anger. Even the most introspective of images emanated vigor.

"Good afternoon, class," she began. "Your first writing assignment is today. I want you to write a two-page essay. I have these photos to stimulate your emotions, or you can dig into your memories, your past. Either way, I want you to find something important from your childhood and write about it."

The class looked around at the photographs, then at each other. Several groaned.

"You might find this a little hard at first. But it will be easy once you get started. Then the story will write itself."

With a little prodding, the class began. Eva Rozen sat immobile. She looked over at Jim. He sat, frowning at the blank display on his dataslate. Erickson walked over to them.

Eva held up her hand and said one word, "Don't."

"I'm sorry?" Erickson said.

"I'm not here to learn to write some little stories. I'm here for science and this is—this is not for me. Thanks anyway."

Eva rolled up her dataslate, turned back and looked at Jim Ecco. She held his eye, paused, and then stood to leave the classroom.

Marta turned to Jim, mouth agape. He shrugged. The class's attention was fixed first on the doll-like student, then on the teacher. Ms. Erickson checked the roster. "Ms. Rozen, maybe you and I can work together for a few minutes and I can help you get started."

"Maybe not," Eva said.

"Ms. Rozen, life is more than facts, figures, and calculations. Your history makes you who you are. You want to be a scientist? Great! Write about why science is important to you. But focus on your feelings. That's what gives scientists inspiration and intuition."

Eva Rozen held the teacher's gaze.

"Ms. Erickson," Eva spoke quietly. "I know what shaped me and it's private. And I don't want to *be* a scientist—I *am* a scientist."

"You're a student and you must do the assignment. I respect your goals but you cannot simply walk out."

"Yes, I can. You like to teach? Fine. Teach. But my history and my feelings are private. I'll be back to class tomorrow. Maybe I'll even stay."

"Ms. Rozen, why is this so upsetting? I don't understand your reaction. What's wrong?"

Eva's reply was measured. She bit off each word. "I have no use for stories."

"You're missing an important part of your education. The arts shape you as a person."

"Think so? That's what writing did for you? Turned you into, what? A thief? Stealing ideas from your students? Go watch someone else bare her soul."

Erickson flushed a deep scarlet then closed her eyes for a moment. "Ms. Rozen," she said deliberately, "I think you're just plain lazy."

"What did you say to me?" Eva's voice somehow managed to be flat and menacing.

Erickson ignored the implied threat. "Do you treat science the way you treat writing? Do you look only at the electrons and ignore the nucleus? Maybe you think that since electrons can form the bonds with other atoms, then who cares about the nucleus?' Is that how science works? You would have done very well in the Inquisition. Your attitude towards the arts seems remarkably close to the attitude of the Inquisitors towards Galileo in 1615."

Now it was Eva's turn to color.

"Oh, I see, Ms. Rozen. You didn't think I'd know anything about science, now did you? Well, stay or go. That's your choice. And I can keep you or flunk you. That's my choice. But the important thing is not the grade, but the kind of person you are. You must understand the building blocks of human nature as surely as you need to study the periodic table of the elements. Won't you please stay?"

Eva stared without expression at the teacher and then

disappeared through the door. The class sat, stunned. A girl who spoke like a woman and who had treated the teacher like a girl? A soft-looking boy who took on the feared Padron? This had become a day to remember.

Eva stood outside the classroom. She hoped Jim would join her. There was something familiar in his bearing. Sad? Angry? Bad memories?

Eva understood all too well. Her memories were still fresh. They travelled with her from half a world away. She catalogued each memory as a voice. Each murmured and spoke and shouted. Mama and Papa and Bare Chest and Doran were shrill, mocking, animate. Even Gergana sat at the Table of Clamorous Voices.

But weren't they were just memories? Ought they not to have paled? Lost color and tone and depth? Weren't the dead supposed to decompose?

06

AN EIGHTEEN INCH JOURNEY

LOS POBLADORES HIGH SCHOOL

EAST LOS ANGELES

2022-2026

Eva marched out of Henna Erickson's classroom, leaving behind two dozen bewildered faces. Jim stood and followed Eva. He turned briefly to the instructor and gave a 'what else can I do?' shrug. He said, "I'll go make sure she's okay," and left.

"Eva," he called once he was in the hallway. "Wait up."

He reached her. They headed across the campus in search of Ringer. Eva thought it odd to have someone follow her without feeling an accompanying sense of danger.

Eva wanted to purge her thoughts of the violent turn her life had taken scant weeks ago, to disgorge the history that Mrs. Erickson wanted her to recall and inhabit. She thought instead of today's events, of this smooth-faced boy next to her and his sudden snarling transformation. She remembered the threatening voice of Padron; it evoked those of Bare Chest and Papa. The memories cascaded, and she heard Gergana and Coombs and every person she'd encountered during her thirteen years of life. For Eva, memory was sound: the din from the Table of Clamorous Voices.

She shook her head to clear the memories. She liked to imagine that she possessed a stage magician's box. Its black lacquered sides were studded with dull iron fasteners and circled by heavy chains and a padlock. With a snap of her stubby fingers, Doran and Bare Chest went into the box. Snap! Henna Erickson. Snap! Mama and Papa. Snap, snap! The box shrank until it fit into her pocket. It never quite disappeared, though, and the Voices were never quite stilled. Mama's whine and Papa's drunken manifestos, Gergana's silly chatter and affectionate lullabies. Doran's grunts and Bare Chest's threats. All of these echoed. The din.

Eva tore herself away from her daydream and turned her attention to Jim. He was frowning. "Bad memories?" she asked.

"I don't want to talk about it." he replied. They walked in silence towards a shady spot on the edge of the school's campus. Ringer was waiting there. She sniffed Eva and wagged her tail. Her ears were relaxed, her tongue hanging down, spatulate. She pressed up against Eva and then returned to Jim's side. Jim brightened.

Eva said, "I have an hour before the Foundation car picks me up. You want to do something?"

"Like what?"

"Something to eat? Anything."

"I don't think so. I need to get Ringer back home."

Eva pushed on. "How about tomorrow? Or the weekend? Let's compare notes. Maybe we can make trouble." She offered a version of what she imagined was a sly smile.

Jim regarded her for a minute. "I liked the way you stood up for yourself this morning. That was pretty cool."

"Okay, I'm cool. You're cool. So...let's do something. Something cool."

"I don't know. I've got to head on home."

"What, somebody keeps track of you? Times your arrival?"

"No, it's not that—"

"Maybe you think I'm not good enough for you?" She turned and stood in front of him, stopping him. She thrust out a clenched jaw. The din from the Table was louder.

Jim held up both hand in a peacemaking gesture. "No, that's not it. Okay, you're different. You're not like anybody I've ever met. You're, what, a scrap over four feet tall? And you were the first to take on those guys. I guess I admire you." He walked several paces, kicking at stones as he went. "Do you want be friends?"

"Friends, huh?" she replied. But the edge was gone. The Table quieted.

Jim sighed. "I could use a friend. Somebody I can trust."

"How do you know you can trust me? You don't know anything about me."

"Ringer trusts you. Let's find her some water and get a soda or something."

"I guess so," she nodded. They walked in silent fellowship towards the nearby diner. Ringer strained forward when she saw their destination, hindquarters shaking from the rapid movement of her tail. Jim led Eva inside to a pair of old-fashioned counter stools. At the base of one, there was a folded blanket with a well-worn depression and a layer of tan hair. Ringer curled up in the depression. The counterman gave Jim a fresh bowl of water for Ringer and served Jim and Eva's sodas, then delivered a small plate of raw burger meat to the dog. Ringer emitted a quiet chuffing sound of approval. The cook was well-trained.

Jim and Eva sat in silence for several minutes. "That writing assignment was weird," Jim said, at last. Eva did not reply. She thought about her life in Sofia, and the last time she had seen Gergana. Eva had kept the scarab brooch she'd never had a chance to give her sister. *No*, Eva thought, I'm not going to spend much time in that class. She reached her hand up and tentatively, touched Jim's shoulder. He turned to her and offered a neutral smile. Her hand fell back to the counter. Jim reached over and squeezed her hand.

"Friends," he said, with a smile as genuine as Coombs, and squeezed her hand again.

The din was gone, the Table was silent. Space opened up at the Table to admit a new member. Jim stood at its head. He exerted a powerful influence, calming the others. In his presence, Eva felt a respite from the din.

❁   ❁   ❁

Marta, Eva, and the Hidden Scholar Foundation car converged at the school's front steps. Marta had a faraway look and Eva asked, "You okay? You look like you've seen a ghost."

"Sort of," Marta said.

"What happened?" asked Eva.

"I took the writing assignment seriously. It brought back some memories."

Eva rolled her eyes. "So, what, you're better than me?"

"Why would you say that, Eva?" Marta sounded surprised.

Eva mimicked her classmate, her voice taking on the singsong, dreamy quality of Marta's reply, "*I took the writing assignment seriously.* Look, what nobody seems to understand is that I don't need stories. I am science," she said, a bit of her old accent spilling back into her speech, a clue to her sudden anger. Eva paused, "So, what did you write about so…seriously?"

Marta stared at Eva before she replied. "I wrote about my parents." She hesitated a few moments and then added quietly, "My mom died a few months ago. She was sick and my dad didn't take it well. I ended up spending the summer with my grandmother. Maybe I have seen a ghost."

Her voice was both testy and sorrowful. Eva looked at her and then reached out and touched Marta's forearm. Today, the gesture was one of solidarity. In time, the gesture would be as much a warning as a cobra's hiss "Here's the Foundation driver," she said.

They got into the car. Marta smiled at the driver, leaned back in her seat, and closed her eyes. It had been a long day. Pain etched a grimace on her face. Eva looked out the window and saw Jim and Ringer.

"Hey, driver," she said. "Pull over. I want to give our friend a ride."

"Sorry, miss, I can only take the students from the Foundation."

"Well, today you can make an exception."

"Sorry, miss," the driver repeated.

Eva threw her door open, forcing the car to stop.

"Miss, please close the door."

Eva ignored the driver and called out to Jim. "Yo, Ecco. You want a ride? The driver says he would be delighted to give you a lift." She drew out the word, looked at the driver, and arranged her mouth into the approximate shape of a smile. Her eyes were hard. She hopped into the front seat, startling the driver, and said, in a near whisper, "Listen. We owe this kid. Somebody tried to jump us this morning and he stopped them. So, just for today, you're going to find a little different route home. Tell your boss there was a detour, something. Help me and one day I'll help you." Then she opened the front passenger door and leaned out again. "Otherwise I drop to the pavement and say that you took off while I was getting in the car."

The driver frowned as if trying to decide which held more danger: her threat or her smile. He pulled over.

"Jim," Eva called. "Get in. We'll give you a ride."

Jim and Ringer got in the car. The driver glared at Eva. She held his gaze until he looked away. Jim looked puzzled, then concerned. A flash lit Eva's eyes. Then they turned opaque, and evicted any attempt to see into her soul. That territory was off-limits.

The riders sat without speaking. Eva was sphinx-like, wrapped in stony silence. The driver kept his eyes fixed ahead. Marta was reengaged in her reverie, eyes closed. Ringer sniffed, hunting for food, and then settled on Jim's lap. She looked back and forth among the friends, lost in their own worlds.

❁ ❁ ❁

Jim and Marta and Eva were inseparable during their freshman year at Los Pobladores. The next year Eva and Marta spent less time together, and Jim divided his time equally between his two friends. As a third year began, he spent more of his time with Marta.

❁ ❁ ❁

Jim Ecco was as skittish as a wren the day that Marta kissed him. When people stood close to him he was anxious, and when Marta moved into his intimate space to embrace him, he was unsettled. His repertoire of responses to members of the two-legged set had been limited to fight, flight, or wary distance, and the movement from impersonal space into a conjoined embrace was a slow journey.

Jim knew that Marta was willing—her pupils widened slightly, she positioned herself to face him squarely, open and inviting. Her head tilted back a fraction, inviting contact. He thought, *It's taken me two years to kiss her,* a moment he'd wanted since meeting her.

Truth be told, she kissed him.

They had met after school on a warm day in early spring. A nearby park offered a few acres of green grass and a hedge of jasmine bushes. The jasmine lent an intoxicating scent and privacy. They'd decided to work together on a homework assignment. Marta had brought a blanket and a small lunch. They'd arranged the blanket and Marta set out a variety of fruits and cheeses, a small loaf of sourdough bread and sparkling water. She'd packed small plates, indistinguishable from bone china, but unbreakable, and two glasses. The place settings were compressible nanoplastics, shape-shifting materials that could organize and reorganize at a molecular level. The glasses collapsed into discs the width of a drinking glass but as thin as a coaster. Gentle pressure on the circumference of the plates allowed them to collapse into equally small discs so that the table settings occupied less space in Marta's bag than a pack of cards.

"You think of everything," Jim said as he took in the small feast.

"I wanted us to have a nice time. Hunger is distracting, don't you think?"

Her words were matter-of-fact, but he heard the warm harmonics of affection in her voice. He was alert, senses aroused. She spoke with a quiet, measured cadence, almost hypnotic, and Jim had to lean in to hear her. As he leaned in, Marta closed the distance between them, an inch, and her movement drew him closer still. Marta's lips parted and she moistened them with the tip of her tongue.

Jim heard blood pound in his ears. His heart sped and every capillary in his body dilated. He felt a flash of warmth like a corona of radiant sunlight. The heat was real but it was all generated from within. Without thinking—finally, without thinking!—Jim closed the tiny gap between them and touched his lips to Marta's.

At first he feared that he'd committed an offense. Perhaps she read his anxiety, for she placed one hand behind his head and held him to her lips. They kissed again. At that moment, Jim Ecco began his life's longest journey, the eighteen-inch passage from his head to his heart.

Seconds or hours later—who could be certain?—Jim and Marta backed up just enough to see each other's faces. Her usual look of curiosity was creased with amusement. "Nice," was all she said, and then pulled him back and kissed him again, slowly. "Like this," she breathed. Jim brushed the plates and food aside and sank to an elbow. She followed in his embrace. He held her in the crook of his arm and played with her hair, stroking and pulling it gently. His hand explored the terrain of her face and he thought he saw something new in her familiar features.

Jim started to speak but Marta placed a finger on his lips. She kissed him again and took his right hand and placed it on her breast. "I will not make love with you today," she whispered. "But I will give myself to you soon. I promise this to you."

He bowed his head in fealty. He removed his hand and kissed her at the soft indentation where her collarbones met. "*Te quiero*, Jim," she breathed. I love you. She held his head against her breast.

Surely the infant Jim had laid his head on his mother's breast. Surely she soothed and comforted him in a loving embrace. He would not have known how to be held and comforted without that experience. But whatever quotient of tenderness had been offered to the infant, he'd existed without it, and the sensation of intimacy with Marta was unfamiliar. They lay together on the blanket, unmoving save for fingers that caressed the outlines of each other's forms. Marta traced his jawline and the soft skin of his neck and then rested her palms on his chest.

"Touch me again," she urged and drew his hand up once more.

He kept her cradled in the crook of his left arm and ran his right hand over the contours of her body, exploring the flat of her stomach and the roll of her hip. She arched her back and pressed herself in closer as he ran his hand over the smooth curve of her buttocks. She breathed into his neck.

"Marta, I feel…funny. No, not funny, but, I don't know…different. Is this what it feels like to be in love?"

She took his hand in hers, and placed both on the center of his chest.

"What does your heart say, *querido*?"

"I don't know. This is all new."

"Your heart knows. Haven't you wanted to kiss me all year? No. No words. Tell me with your heart."

So he kissed her again, now at the corners of her mouth, on each lip and then openmouthed and urgent. His thoughts stilled, replaced by the need to possess and be possessed, to draw her in, to find a calm surcease of anger.

Four days later Marta fulfilled her promise. She gave herself, took his strength in exchange, and passed into womanhood. Jim discovered a still place within himself where turmoil paid obeisance to the gentle parts of his being.

There was no school and the house would be his for the day. He spent the morning cleaning his room, checking for dog hair, pacing and then cleaning again. Marta arrived. She wandered through Jim's home, looking at the photos on the refrigerator, the art on the walls. Ringer kept to her side. When Marta sat at a dining-room chair, Ringer placed her head on the girl's lap. Jim smiled and said, "She beat me to it."

They laughed and stood and embraced and kissed. Marta laid her head on his chest and held him close to her. Together, they swayed to an inaudible rhythm.

"Would you like to make love to me?" she asked.

Jim said nothing. He took her hand and kissed each of her fingers and then led her to his room. They undressed each other in self-conscious wonderment, and handled each piece of clothing with the reverence of a pilgrim touching a holy relic. Jim sank to his knees before her and pressed his head to her stomach. He breathed in deeply, and then sank lower to kiss the gnarled joints of her left leg. She gasped and started to pull away but Jim held her fast, as she had held him four days earlier. He pressed his cheek to her calf and then kissed her feet. She allowed herself to sink onto his bed. She reached to pull back the sheets. They might as well have been cemented in place, they were tucked in so tightly, and they laughed as they struggled to free the linens.

Jim grazed his hands along her legs and paused at the plain dark triangle that held such awe and mystery. He traced the concave line of her ribs, around her breasts and up to her face again, holding her head immobile while he kissed her again and again.

She lifted her legs and placed them on the bed. Jim supported himself above her and allowed her to caress his chest and hips. She reached down and took him in her hands. One moment he was above her, separate, and the next moment he was inside her. They were fused. They kept their eyes open and marveled at the sight of one another. Then they were engulfed in passion.

Later, they lay entwined. Each time Jim started to speak, Marta put her mouth over his mouth to stop him, although she did permit him to profess his love for her. Repeatedly.

Much later, Marta broke the silence. "Why did you wait so long?" she asked.

The next day at school, Eva stopped and looked first at Jim, then at Marta. Pain, then anger flashed across her eyes, almost too brief to notice. Then she grinned.

"About damned time," she said, and lapsed into stony silence for the rest of the day. It was difficult to think when the din from the Voices at Table rose in deafening ridicule.

❀   ❀   ❀

Throughout high school, Jim, Marta, and Eva, friends by exclusion as much as by attraction, were protective of one another even as they quarreled. Marta and Jim sometimes fought, always over Jim's temper. His anger was hard for her. Eva's insults were mingled with affection. She kept Marta close, but always at arm's length, as if the act of embracing her would be painful.

Jim supplied the minimum effort to pass his classes and remain enrolled. He continued to study people, teasing out their secrets, a talent that often proved more curse than blessing, a gift that cleaved him from, rather than bound him to, members of his own species.

Eva and Marta, drawn to science since childhood, were accepted at Yale, Tufts, and Harvard. They chose Harvard College, for its medical school and its Center for Nanoscale Systems. The Hidden Scholar Foundation continued to fund their education.

And the three friends who shared different but difficult childhoods, three friends thrown together by chance, three who would share an orbit travelled to another part of the country and another chapter in their lives.

07

WALKING WITH JURICÁN

BOSTON

OCTOBER, 2029

Okay, Jim thinks, she was old. She had cataracts. So what? She didn't need to read a dataslate. Her hearing had largely faded. But she registered the clink of a spoon at mealtimes. She smelled bad… so what?

Ringer had been Jim's friend and familiar. The move from southern California to Boston had been hard enough on Jim, but the freezing weather and endless grey skies seemed to drain life from Ringer. Her filmy eyes implored, a whine in her voice chided, *Make it warm*! Her every arthritic step made Jim ache. A stab of pain shot through him each time she fell.

A boy and his dog. Ringer was eleven, and then—no more.

Tips from, "Coping With The Loss of Your Dog"
It is normal to feel angry

"God help me, Marta, I'm about to explode."

"Shhh…querido. Just let me hold you for a while."

"YOU DON'T UNDERSTAND DAMMITTHATWAS RINGER."

It is normal to feel depressed

"Jim, it's been a week. Are you going out? What about your job?"

"Leave me alone."

"It's not just you. I can't ignore my classes. Harvard is harder than Los Pobladores.

"I'm not keeping you from your studies."

"How can I concentrate on school when I'm worried about you? I have to go to class. Jim? Jim?"

The most important step in your recovery is to express your feelings in a way that suits you best

Night shrouded Boston Common. Dark figures slid in and out of the park's shadows: drug dealers, prostitutes, muggers—and a hunter. Jim Ecco moved silently. His practiced eyes counted the park's denizens. He wanted to find two or three, young enough to be a challenge, but not so young as to be exculpable.

He crouched by the Soldier and Sailor Monument. It topped a small rise and gave him a view of much of the park's fifty acres. Over the centuries, the Common had hosted soldiers, protests, and recreation. It accommodated criminals as graciously as upright citizens. And on a crisp fall evening, as the fires of his rage and the anguish of his guilt consumed him, Jim Ecco prepared to approach three of these habitués.

He moved away from the monument's bas relief, stepping with care. No twigs snapped, no leaves crackled. He raced across open ground, crouching low, moving with the cover of trees. He kept his attention wide, sensing for danger, for intrusion, for anything that might come between him and his prey.

He'd marked his quarry the night before. Now he drew into himself, presenting the smallest possible profile as he closed the distance to the trio. They would get an opportunity to leave him alone, although the assaults he'd observed at their hands argued against a peaceful interchange. They would set on him and Jim would respond. Violence would be his catharsis. He could purge the feelings of helplessness and frustration that he'd felt at the hands of his father, the grief he felt at the loss of Ringer.

Twenty yards, fifteen, and still no sign of recognition from them. Five yards. It was time. His posture changed, he stood upright, and shed his stealth. One of the three looked up at him and then nudged his two companions. They fanned out around him.

Jim relaxed. His eyes became unfocused as he took in the *gestalt* of the night. His prey came closer. He'd evaded his father with ease, and Padron, at school. These three would be no harder. They would strike first, but he would land the telling blows.

He heard a rustling to his right. Two more figures came into view, two women. One stood no taller than a child. The other limped in obvious discomfort. *Eva? Marta? What the—?*

Time sped. The women turned to Jim's presumptive victims. The small one looked at the closest of the three men and grinned. There was no humor in her smile.

"Boys, it's been fun, but you were just leaving."

Marta Cruz lumbered over to Jim. "Don't do this, querido. You cannot fight and stay whole. Juricán walks with you tonight. He will destroy you. Come back to me."

Suddenly, the man closest to Eva Rozen lunged at her. One moment she was in front of him, and then she seemed to vanish. The man flailed wildly, bewildered at her sudden disappearance.

She reappeared behind him. "I said, time to go. Here—buy yourselves a drink somewhere. Anywhere. But not here." She held out a bill.

He lunged again, and once more she disappeared and reappeared beside him. "Okay, two drinks," and now there were two bills in her hand.

The three men looked confused as Eva winkled in and out of view. Marta spoke to them, softly. "I have something better for you than violence. I bring you life. Leave now and you will have a much better evening." She withdrew a handful of fine powder from a small leather pouch she wore around her neck. Marta blew on the loess in her palm and it enveloped two of the men. As they breathed in the airborne particles, their features relaxed.

Marta continued in a soothing voice. "That is the seeds of the cojobana tree. You will have a vision. You should heed it well, as it is a gift from Yocahu."

Eva's voice was a sharp contrast to Marta's gentle words. "Well, boys, which is going to be? The lady," she pointed to Marta, "or the tiger? That would be me. You know what? I think her gods are nutty, but you'll like her approach a whole lot better than mine." Eva vanished again, popping up next to the trio's ringleader. "Time to decide. And, just so you know, I have something different up my sleeve and I promise you won't like it."

She had money in her right hand and a small squeeze bottle in her left. The smell of peppers drifted in the crisp air. The leader of the three looked at her and shook his head, trying to clear his thoughts. He looked at the money and snatched it from her hand, turned and strode towards Tremont Street and away from this strange tableau. The other two remained, rooted to the spot, rapt in the beginnings of a vision.

"Come on, children, time to leave before we have any more company," said Eva.

Jim Ecco recovered from his surprise. He looked at Marta. "What are you doing? Are you crazy? You could have been hurt. This is not your fight."

"It is not your fight, either."

"I needed to do this. And those three are no good. Putting them out of commission for a while would be a blessing for everyone." He paused, confused by Marta and Eva's sudden appearance. "How did you get here, anyway?"

Eva said, "You aren't hard to follow. Weren't too hard last night, either."

Marta grasped his shoulders. "Look at me, querido. Look at me!" Her eyes shone. "Juricán is powerful and gives you strength but he will take it away. You cannot walk with him and survive. He consumes his followers. The hurricane is too great for you and your anger feeds it. Tonight you need grace."

"Marta, I don't know about any of that. I just know I can't stand the way I feel."

"Then you need to find a new way to feel."

Marta turned to her companion. "Thank you, Eva. No one needed to be hurt, and no one was."

"The night's still young."

Jim was silent for a few moments then turned to Eva. "How did you disappear like that?"

"Smartwool," she said. "It's the latest fashion in stealth wear. It absorbs odors. Sheds water and stains. And the nanothreads are glass and plastic. They change colors instantly. I had them programmed to reflect the colors around me so that I appear invisible.

Military's been using this stuff for years now. Neat, huh? Bet you wish you had a snappy little outfit like this."

Marta spoke. "Uh, guys, I think it's time we get out of here. We're beginning to draw some attention."

Jim and Eva looked around. "Nothing here we can't deal with," Jim said.

"You're wrong. There are three things we can't deal with," Marta said. "We need to leave—now."

"What three things?" asked Jim as they began walking. Dried leaves crunched under their feet as they crossed the park to the busy street. There was just enough light from Tremont Street's shops and streetlights to cast shadows across their path. The hum of traffic and voices on the sidewalk grew louder as they approached the park's end. Jim heard something different in Marta's voice, a new vitality that animated her words. He tuned out all of the night's sounds to concentrate on Marta.

"Well, first of all, there's a curfew and we're violating it. You two might feel immune, but I don't."

"Okay. What else?" asked Jim.

"This is taking a lot out of me. My legs hurt. And I have a ton of schoolwork. Premed's tough, and I'm swamped. I've got an organic chemistry project due. Eva's the chem whiz. It's no picnic for me."

"Oh, jeez, I'm sorry. Right. Let's go," Jim said.

They walked across the Common, headed for the bright of the Park Street station of the underground transit line.

"You said there were three things. What's the third?" he asked.

Marta beamed. "I'm pregnant."

The assault came, vicious and unexpected, just fifty feet from the lights, clamor, and activity of Tremont Street. Hooting in surprise and congratulations, the three friends failed to hear running footsteps behind them. Then Eva tumbled to the ground.

At his trial, Eva's assailant Brian Coogan would testify under a forensic dosage of TrueSpeak that he wanted to reach her before she was out of the park. He was furious. Eva had embarrassed him and the money she'd given him was gone. He would state that he thought that somehow she'd taken the money back. (She had.) He would tell the court that when Eva fell forward, he believed that his knife had penetrated her heart.

Eva survived Coogan's knife attack with deep bruises but no other injury. Her snappy little outfit included magnetic shearing fluid within woven carbon fibers. Iron nanoparticles, suspended in the fluid, solidified in a microsecond when subjected to stress. It was an old technology, first deployed with mixed results in the early years of the U.S. wars with Iraq and Afghanistan. A quarter century of development, and the armor was more effective than Kevlar or spider silk. The principal customers for this class of smart textiles were police and military organizations…and Eva Rozen.

Perhaps Juricán did indeed walk with Jim Ecco that evening. When Eva fell, Jim reacted instantly. He pushed Marta out of the way and turned to Coogan. The attack reanimated Jim's rage. His vision narrowed uncharacteristically. He missed seeing Eva roll on the ground, bloodless and then struggle to her feet. He missed the sight of two Boston police officers running to their aid.

Jim continued his turn and lashed out with a kick that landed on the side of Coogan's left knee, just below the patella. Coogan crumpled in pain. In their statements, the patrolmen would report that they ran to protect the three victims when they saw Coogan fall and hold his left leg. They saw Eva begin to rise. Then they saw Jim turn and stomp on Coogan's right foot.

Two things happened. The fifth metatarsal bone at the base of Coogan's little toe cracked apart, producing a crippling break called a Jones fracture. Despite nanoputty surgeons applied to the bone, Coogan would wear treatment cloth for three weeks. The second result was Jim Ecco's own arrest: assault and battery with a deadly weapon. The weapon was, in the language of his indictment, a shod foot.

Within the space of a minute, Jim Ecco found that he would be a father...and a felon.

08

 TWO VERDICTS

Sean Doyle rose for his opening statement on behalf of the People of Massachusetts. His navy pinstriped suit clung to the contours of his six-foot frame and fell from broad shoulders to hug a trim waist. Doyle's red and blue striped club tie fastened itself into a perfect Windsor knot and bisected an unblemished white shirt, stopping precisely at the top of his beltline. Doyle's head was adorned by thick curly blond hair, augmented eyesight, and a surgically-crafted cleft chin. Here stood a man the jury could trust on sight, a man who could lead them to discover justice on a spring morning in the Suffolk County Municipal Court in Boston, Massachusetts.

The People's representative addressed the presiding judge, the

Honorable Chris McClincy. Doyle stated his name for the record. He was given leave to begin his opening statement. He stood without notes and faced the jury. These six men and women took in his self-confidence and beamed with reflected pride. Ordinary men and women, working people, retirees, salt of the earth now stood shoulder to shoulder with this mighty champion. They would join him in a pact to protect the Commonwealth, to honor the Law, and to send that son-of-a-bitch in the defendant's chair to whatever dark corner of the penal system he deserved, God have mercy on his contemptible soul.

Doyle spoke. Twelve ears edged forward to listen. Twelve eyes focused to watch their prophet. Six hearts readied to be blackened. "Ladies and Gentlemen of the Jury, this is Case 260093, the Commonwealth v. James Bradley Ecco. The defendant is charged with having committed an intentional assault and battery by means of a dangerous weapon, his shod foot. Section 15(b) of chapter 265 of our General Laws makes this act unlawful."

Assistant District Attorney Sean Doyle was ambitious. Step One in the Doyle Plan: District Attorney. Then Attorney General for the Commonwealth en route to Congress. Or to the governor's office. Then, who knows? Presidents have been created from similar pedigrees.

Seniority granted him the ability to choose cases that bolstered his conviction rate. This case was a prosecutor's dream: an injured victim and two unimpeachable witnesses, sworn officers of the Boston Police Department. Doyle was cultivating a tough-on-crime reputation and had refused the defense offer of a plea bargain.

Doyle's pinstriped suit followed him faithfully as he paced behind his lectern. Two steps to the left; two steps to the right, the pendulum in God's hypnotic metronome. Left, Right. Tock, tock. Left, right. Tock, tock. Gazing into an unseen vision, he became a Loa-possessed demon, Michael the Archangel and Skadi, the Norse paladin of Justice, Vengeance, and Righteous Anger. Turning to his jury, *his* jury, he was holy, fair, and humble. "Ladies and Gentlemen, there are rules that I must follow, and you must hold me accountable."

These words elevated the panel and he was rewarded with six faces set with grave dignity. The prosecutor paused before them, arms outstretched, an Old Testament prophet. If he could embrace them, he would.

"The Commonwealth must prove beyond a reasonable doubt that the defendant intended to touch the victim with a dangerous weapon. You must require, no, demand proof from me that the defendant intended the touching to occur, that it was no mere accident." The jury leaned forward as if to lay their hands on the trailing hem of his pinstriped garment.

"Let's talk about what the law does *not* require. First, the Commonwealth is not required to prove that the defendant intended to injure Mr. Coogan—only that he did cause harm. Next, it is true that the victim is a criminal, but the victim is not on trial. Finally, it is also true that Mr. Ecco at first acted in self-defense. But he did not stop when his victim was helpless and that, ladies and gentlemen, is Mr. Ecco's crime."

Doyle filled the jury with glowing comprehension. They were clay and he would breathe life into them, endue them with holy purpose: to perform their duty to convict.

"What about the 'dangerous weapon'? A shoe? Not a gun, knife, or club? Well, the law says that even an innocent item is a weapon if you use it in a dangerous way. Your old-fashioned pencil is an innocent item if you are writing a grocery list. But if you poke me in the eye, then it a dangerous weapon. When the defendant used his shoe, not to walk away from a helpless man, but to stomp on a helpless man, then his shoe became a deadly weapon."

Doyle paused, seventy-two inches of moral outrage, and pointed. "The defendant may look harmless, but he has a vicious temper. He committed a crime in plain sight of two police officers. This case is simple and the People of Massachusetts depend on you to do your duty."

Sean Doyle looked at the jury, thanked each member. They watched in awe as he and his pinstriped suit returned to the prosecutor's table.

Two women waited outside the courtroom, witnesses in the case and unable to attend the proceedings until called to testify. Marta fretted. Eva seemed detached, even bored.

"How can you take this so lightly?" Marta asked Eva.

"Why worry?"

"Why worry? Because Jim drew a prosecutor with political motives and a judge who favors prosecutors. Lord knows what the jury will do," said Marta.

"Can't worry about what you can't control. Just control what you can." Eva touched a device wrapped around her forearm and began to subvocalize.

Marta stared at her friend in goggle-faced astonishment. "Oh, my god. Is that what I think it is?"

"What do you think it is?"

"Where in the world did you get a datasleeve? They cost a fortune! I don't understand…did the Foundation buy it for you?"

"Nope."

"How does a college junior end up with a datasleeve?"

"I deserved it."

"Right," drawled Marta, but then she was given over to

curiosity. "Is it true that the nanoprocessors eliminate all of the heat that a dataslate generates?"

"See for yourself."

Eva held up her wrist and Marta examined the device. It was nanotextile, wearable electronics, components thousands of times smaller than human hair. The sleeve was about the thickness of flannel. It packed the computing punch of a massively-parallel mainframe from an earlier semiconductor generation.

"How does it work?" asked Marta. "I thought they had to communicate with datapillars. No, don't tell me you have a pillar. Isn't that a little beyond even you?"

"They do have to sync with a datapillar," said Eva.

"Tell me what I'm missing," said Marta. "Only governmental agencies have pillar-and-sleeve technology."

"They're being deployed commercially now, too. In a few years, everyone will have them. I'm just a bit ahead," said Eva.

"So, do you have a pillar?" Marta pressed.

"Not yet."

"Then what good is it?"

"Well, you can sync with anyone else's pillar if you know how," Eva grinned. "Anyway, sleeves will be on the market for the public in a year or two." *By that time,* Eva thought, *I'll know how to jack anyone's sleeve.*

"How did you get it? This is incredible. Can I try it?" asked Marta.

Eva held Marta's gaze. "Where are the other witnesses? Where's that bastard, Coogan?" Without turning her attention from Marta, without looking around the courtroom hallway, she observed, "I don't see the prosecution witnesses. You suppose that's a good sign?"

Inside the courtroom, Sean Doyle suppressed a grin as the public defender rose for his opening statement. Three members of the jury crossed their arms as if to wall themselves off from the hapless attorney. Another two glared. The final juror's eyes were fixed on her feet, as if she'd found webbing where she expected to see sensible shoes. None made eye contact with the overmatched defense counsel. He reminded the jury that his client was innocent until proven guilty, that his client had been protecting himself and his friends and had every reason to fear for his life. The defense counsel asked for their patience and skulked back to his seat.

Judge McClincy turned to the prosecutor. "Mr. Doyle, are you ready to call your first witness?"

"I am, Your Honor."

"Then please proceed."

Doyle rose, nodded to the magistrate, looked at the jury and in a clear confident voice, began his prosecution. "The People call Brian Coogan."

Every eye in the courtroom turned to the secured door from which Coogan would emerge. The quiet lasted for thirty long seconds. The witness did not appear. Judge McClincy turned an expressionless glance to Doyle asked the prosecutor again for his witness.

"Judge, he is on the witness list. Would Your Honor direct the bailiff to locate him?"

"Mr. Doyle, I expect attorneys to be prepared in my court. Please call another witness while the bailiff finds your Mr. Coogan." McClincy nodded to a burly, uniformed man who nodded back and entered a secured holding area attached to the courtroom.

"Thank you, Judge. In that case, the People call William

Stevens." Stevens was one of the two police officers who witnessed Coogan's attack on Rozen and Ecco's subsequent assault. Eyes turned to a door at the rear of the courtroom, expecting to see the officer appear. The door stayed closed.

The bailiff reappeared and told the judge. "Mr. Coogan did not arrive. The defense witnesses are outside but none of the prosecution witnesses are here."

Doyle's ruddy cheeks drained of color. He looked up at Judge McClincy and spoke, "Judge, all of my witnesses are supposed to be here. There must be a minor mix-up. May I have a brief recess to get this straightened out?"

"Five minutes. Please have your witnesses when we reconvene. The jury will remain."

McClincy gaveled the session to a halt and then returned to his chamber, but not before offering a dark look at the prosecutor whose temerity interfered with His Honor's schedule. Three hundred seconds later Judge McClincy returned. The bailiff called the session back to order. McClincy turned to the prosecutor and asked, "Are you ready to continue, Mr. Doyle?"

"Your Honor, may I approach?"

Permission granted. Doyle and the public defender rose and approached the bench. Doyle was subdued. "Judge, the witnesses appear to have been sent to a different court. Somehow a change of venue order was entered in error and my witnesses were sent to Franklin County."

"Not good, Mr. Doyle. How did a Suffolk County prosecutor manage to send three witnesses to a court ninety-one miles from here?"

"I don't know, Judge. We're still looking into the matter. May I ask for a continuance?"

"Do you at least have statements from the witnesses?" asked the judge.

"Of course. May I present those?"

"If you don't, I'm guessing that the defense here will ask for a dismissal."

Doyle walked to the prosecutor's table and conferred with the second chair, his assistant. Doyle's demeanor changed from confident to frantic as he subvocalized commands to his datasleeve. He checked an index of documents and blanched.

Doyle's pinstriped suit pulled the reluctant prosecutor to face the bench. The perfect Windsor knot was askew. Once again he asked permission to approach, and once again the public defender accompanied Doyle to speak with His Honor.

"Judge, it appears that the witness statements are not available."

"Not available now…or not available ever?"

"The written statements were destroyed in error, Your Honor, and the backup is gone. If I could have a postponement…"

The public defender showed his first signs of life. He was overmatched, but understood the most basic tenet of criminal defense. No evidence? No victim? No witness? No case. He called for a dismissal.

"I'm inclined to grant the defense motion,"

"Your Honor! The defense request is outrageous. The People have spent days preparing for this case. If you can give us a continuance, we can reassemble the testimony and get the witnesses into court."

McClincy glared at Doyle. "As I recall, you refused a plea bargain offer from the defense. No counteroffer, either. Do I remember correctly?"

"Yes, Judge."

"I don't like trials. They interfere with justice. Too inefficient.

And now you want a continuance? You reap what you sow, Mr. Doyle."

"Your Honor. Mr. Coogan deserves justice."

"Chambers, both of you," McClincy snapped. He looked up and addressed the court. "We're going to take a recess here for, let's say, fifteen minutes. Bailiff, please escort the jury out. Ladies and gentlemen, remember not to discuss the case. We'll call you back when we're ready to reconvene." The jury turned a piteous glance on Doyle, their fallen hero. They rent their garments, covered themselves in ash, and wailed in grief.

McClincy turned to the defense. "Mr. Ecco, I'm going to have a little chat with your attorney and Mr. Doyle. Would you be kind enough to join us?"

Without waiting for an answer, McClincy walked behind his bench and through a door leading to his chambers. Dark wood-paneled shelves accommodated photographs of His Honor's family. The walls were papered in a *trompe l'oeil* image of books—codes, statutes, ordinances, decisions—an artist's notion of a jurist's sanctuary during the gaudy age of paper. Maroon pile carpet finished the effect, a gentlemen's club. McClincy gestured to a pair of adjoining leather chairs for the attorneys. Jim looked around for a chair. There wasn't one. He stood. A court police officer stood at his elbow, ready to pounce, ready to protect the assembled officials of the court.

"Mr. Ecco, there are a few details that puzzle me. Help me make sense of them."

"Your Honor?"

"Mr. Ecco, our SETS system is about ten years old. That's the Safe Evidence and Testimony Storage system. Evidence is stockpiled electronically. It removes the delays in handling evidence and documents. That ensures that you get a speedy trial. SETS hasn't lost one byte of data in all of the years it's been in existence, at least that I know of. Mr. Doyle, have you lost any evidence before today?"

"None, Your Honor."

"Mr. Goodrich, are you aware of any lapses in SETS?" McClincy asked the public defender.

"No, Judge."

"So, Mr. Ecco, the likelihood that the People's electronic and hard copy statements were destroyed *and* the witnesses were sent elsewhere is a bit lower than the likelihood that Mr. Doyle here will show up in Bermuda shorts instead of his usual pinstripes. Add the fact that the defense evidence is intact and your witnesses are here. Would you care to speculate on the odds?"

"If you think I had anything to do with it, you're wrong. There are smarter people accused of bigger crimes who have more ability to tamper with evidence."

"I agree, or we would be having a different conversation. Still there are just too many coincidences, and that bothers me. Mr. Doyle, has your office figured out what happened?"

"Judge, it appears that there was a similar case and a defendant with almost the same name, Jaime Eccles. He was charged with assault and battery with a deadly weapon, same as Mr. Ecco. That case was dismissed. Because of the dismissal, the witness statements were to be deleted. But that order was applied to this case in error."

"What about your witnesses? The policemen? The victim?" asked the judge.

"I don't know why they were sent to Franklin County. There was no change of venue in the Eccles case."

"Can you at least tell me how it happened?"

"As far as I can tell, the instructions were issued by the datapillar. That's all I know."

Jim's eyes widened at the mention of the datapillar. The

attorneys were silent, pensive, waiting. The judge looked around his chambers as if the solution would be found on the fanciful walls. He fixed his gaze on Jim and drummed the fingers of his right hand on his desk, slow and light at first. The cadence moved from andante to moderato to a vigorous presto. Finally, agitato, and crescendo. Then silence.

"Mr. Ecco, every night, I review the next day's docket. Your case was straightforward. The victim attacked your friend and you defended her. Fine. But when her assailant was immobilized, you kicked him. That's the assault, which was witnessed by two policemen. I was prepared to sentence you to two and a half years in the House of Corrections. The sentence is severe, but you've been in court before, haven't you?"

"I have a clean record, Your Honor."

"Mr. Ecco, that was a 'yes' or 'no 'question. Again. You have appeared in court on three occasions, including this one, correct? And each of your visits came about because you have a bit of a temper, yes?"

"Uh, yes, sir."

"And, in each case, you…how shall I put this? You *whaled* on your adversary. I believe that's the term I'm looking for. Is that right Sean? Whaled?" Doyle nodded. Whatever His Honor proposed was right.

"Your Honor, I defended myself."

McClincy took in a deep breath. His cheeks puffed as he exhaled through pursed lips. He spoke slowly, as if addressing a child. "Mr. Ecco, you have a temper. It is out of control. You have a habit of breaking bones. If any of the People's evidence were intact, you would be the County's guest for the next few years. I will not let you off because of a mix-up.

"Here is your choice. I can grant Mr. Doyle's request for a

continuance, or I will allow your attorney and Mr. Doyle to reach an agreement. Mr. Doyle, how about disorderly conduct? Keep your nice conviction rate intact?"

"The People would accept that, Judge."

"Defense?"

"We would accept disorderly, Your Honor, and request probation."

"Mr. Doyle?"

"No, Your Honor, Mr. Ecco committed a grievous crime. He—"

"Oh, knock it off, Sean. This isn't the courtroom and I'm not a jury. You wanted a conviction. Now you've got one. I want him on probation so I can force some changes on him."

"Community service, Judge?" requested Doyle.

"My thoughts exactly."

The magistrate turned to the defendant. "Mr. Ecco, if you plead guilty to disorderly, you'll have a misdemeanor criminal record. You'll be on probation, which means you'll be under the court's supervision. But it also means that you'll walk out of here today."

"I understand, your Honor."

"I'm also going to order you to perform 120 hours of community service."

"Yes, Your Honor." Jim exhaled a deep sigh of relief.

"Mr. Ecco, we're not quite done. As part of the plea deal, I am ordering you to take mood blocks or to enroll in anger management counseling. I am serious and I do not lose track of witnesses or defendents. Do I make myself eminently clear?"

"Yes, Your Honor."

"I'm placing you under the court's supervision for the next two years. If I see you in my court or hear that you are visiting any of my colleagues, you will serve time. Is that understood?"

"Yes, Your Honor."

"Good. Mr. Ecco, please make some arrangements with a physician or a mental health worker. You have ten days to present a treatment plan to this court. Do you have any questions?"

Ecco thought for a moment. His lips parted to reveal a slight smile. "Yes, Your Honor."

"What is it?" McClincy asked, annoyed.

Ecco drew in a breath. "Your Honor, will you marry me?"

"What!"

"I said, 'Will you marry me?'"

McClincy stared, disbelief painted across his face. Doyle winced, and then smirked. McClincy spoke with quiet menace. "Do you think you're funny? Are you a comedian now?"

"No, Your Honor, I'm sorry." Jim raised both hands in a stop gesture. The bailiff grabbed the back of Jim's shirt and pulled him back a step. He continued, "I didn't mean, 'Will you marry me' as in, 'Will you be my spouse.' Please. My girlfriend is sitting outside the courtroom. She's pregnant with our child. We put off getting married because we thought I'd be in prison. I want to do the right thing for her and for our baby. That's what I meant. I should have said, 'will you marry us.' My trial was supposed to last the morning, and now it's over. So, there's time. We've got a friend here for a witness, and, well, you're a judge."

"Do you have a wedding license?" asked McClincy.

"Not yet, sir, but the clerk's office is just down the hall. I'm serious, Your Honor. This trial has been hanging over our heads for months, and every day I've thought about what I did and regretted hurting Mr. Coogan. I'm turning my life around and I want to be there for my family. Please, help me."

McClincy shook his head slowly. "Well, I have managed to dispose of two-hour trial in less than thirty minutes, so if your lady wants to be married in a criminal courtroom, then be back here with a license and I'll marry you after the next case."

Outside the courtroom, the three friends celebrated. Marta threw her arms around Jim as far as her gravid belly would allow and then winced. Impending motherhood buoyed her spirits but burdened her legs. Eva appeared indifferent.

Jim hugged each of the women who had supported him during his ordeal. To Eva, he whispered, "Neat trick. Thank you."

"I'm sure I don't know what you mean."

"Only you could have done it. How can I ever thank you?"

"Take me with you the next time you go hunting."

"Eva, there won't be a next time," and he explained what the judge ordered. "There is one other thing," said Jim. He turned to Eva. She saw a look of entreaty in his eyes. "Marta and I need to stay here for another hour or so. There's one more legal matter and I need your help."

"What's going on?" asked Marta.

Jim's knelt before her and took both of her hands in his. She colored. There was a small cluster of people in the hallway, attorneys hammering out deals, fates being cast. Their voices grew quiet as their attention was drawn to the scene unfolding nearby.

"Marta, you've been my rock for eight years. We have a future in front of us now and I want the best life for our baby. The judge agreed to marry us. To each other, that is."

No one spoke. No one breathed.

"I'm asking you to marry me," Jim said.

The hallway was as silent as a library at midnight.

"All we have to do is walk down the hall and get a license. Marta, I don't know if this is the wedding chapel of your dreams, and there's no wedding veil or ring bearer and all of that. But I want to marry you, Marta. Right here and right now. Will you be my wife?"

"Jim, I don't understand."

The onlookers craned their necks to hear.

"The judge agreed to perform a marriage ceremony after his next case if we get a wedding license. I know this is sudden, but we've talked about it. I want to marry you, Marta." He placed his cheek against Marta's swollen belly and looked up at her. "Let's be a family."

Marta Cruz burst into tears. The attorneys and their clients in the hallway looked at the bride-to-be. Marta looked down at the man kneeling at her feet. She pressed his face into her belly, as if to make the moment include their child.

"Oh, Jim. Yes. Yes, I will marry you."

Every person in the hallway but one started to applaud.

"But there is a condition," said Marta.

Movement stopped again.

"Stand up and look at me."

Eva frowned briefly and watched Jim rise uncertainly to his feet.

"I will marry you, Jim Ecco, and I will be with you for better or for worse. But, the judge speaks for me as well. You have to learn to calm yourself. I won't have our baby nursing on your temper. I know that anger will always be a part of you, but you must become its master, not its servant. If you will promise me that, then, yes, I will marry you, right here and right now." Her eyes began to moisten, as did Jim's.

Eva's eyes were dry. They widened and narrowed. Something tugged at her, an inchoate sense of foreboding, of loss. She shook her head to clear her thoughts. Back in control, she sighed impatiently. "Jeez, guys, this is a wedding, not a funeral. Hey, lovebirds—come on! We've got to get you to the clerk's office."

The onlookers cheered as the three friends ran up the hall. The line at the clerk's window was long, but they had Eva. In less than a minute, they were at the front, and peering through security glass at a clerk.

"Huh. I didn't know that wedding licenses were all that risky," Eva muttered, touching the thick glass.

The clerk chuckled. "Nope. But fishing licenses? Now that's another story."

When the couple returned, Judge McClincy examined the license. He smiled and then retreated to his chambers. He removed a spray of daffodils from a vase on his desk and returned to the front of his bench to stand with the couple. He presented the flowers to Marta and asked, "Will this do for a bridal bouquet?"

She blushed and beamed.

He pointed to Eva. "Are you the maid of honor?"

"Best man. Jim, got a ring?"

"In fact I do." He reached into his pocket and took out a small, cloth-covered box and handed it to Eva. She stared down at the ring. In a low voice Jim said, "Don't lose it," he smiled, "or make it disappear."

Jim turned back to the judge. "Your Honor, we appreciate this. I'm not going to let Marta down—or you. But Marta and Eva have to get to their classes. Can you make this brief?"

Judge McClincy froze. He cocked his head and looked at Jim for a moment, and shrugged. "Okay, you're the boss." He turned to Marta. "You look beautiful, my dear. Do you, Marta Maria Cruz, take James Ecco to be your lawful husband?" She nodded, fighting to control tears, and barely spoke her affirmation.

"And you, James Bradley Ecco, do you take Marta Cruz to be your lawful wife?"

"I do."

"Then by the power vested in me by the Commonwealth of Massachusetts, I now pronounce you husband and wife. Mr. Ecco, you may kiss your bride."

"That's it?" Jim asked.

"You said, 'Make it brief.'" Then, under his breath, "Gotcha."

That evening Marta made a celebratory dinner—a simple student meal of pasta. Eva placed two red rose stems on the table. Jim produced a bottle of student-priced Chianti to toast the future but the plonk went largely untouched. Marta stuck with club soda—"the baby, you know." Jim drank sparingly to a new future. Eva clinked a glass to the others in a toast to friendship, but set the alcohol back on the table, untouched.

The celebration marked a turning point for the three friends. It was the end of Jim's legal troubles and the beginning of his path of self-control. For the first time since the death of Marta's mother and the loss of her father's attention, she concentrated on her own needs. She had made an unequivocal demand of Jim, to abandon his anger, and was rewarded by his efforts to mature.

Eva felt conflicted. When she shepherded Marta and Jim through the clerk's office for a wedding license, she felt a kinship with them, a connection to others that was new to her. True, she had loved Gergana, but as a child loves a caregiver, not as a friend bound by voluntary allegiance. The antiquarian Coombs had offered friendship that transcended the difference in their ages. Eva accepted his benevolence, but not the mutual obligations of amity.

Her growing affinity for Marta and for Jim animated something within Eva. Platonic love, unselfish giving, unrewarded sacrifice—these could nurture her. The desiccated yearning for human connection began to bloom. She felt peace and saw that it was good.

But the Voices from the Table of Clamorous Voices counseled

otherwise. They dripped poison in her ear, an insistent voice with Iago-like guidance. *Something's wrong,* they warned, *here is danger.* Eight years earlier, Gergana's murder had robbed Eva of a model of selfless love and the Table of Clamorous Voices was born. Scant weeks later, Jim Ecco said, "Let's be friends." He offered loyalty without expectation of romantic payment, faith without reservation. He stood at the head of the Table. He became a moderating voice.

Throughout their years together, Eva balanced the fear-driven impulses of the Table with the kindness of Jim's friendship and Marta's tolerant, if sometimes caustic, acceptance. But when Eva pulled them down a crowded hallway to the County Clerk's office, she helped Jim move into Marta's orbit. Eva did not understand the chemistry of human relations, that Jim could be shared by two people, much as an electron is shared by two atoms in the formation of a stable molecule. The Table of Clamorous Voices spoke, and it warned, "You have been eclipsed by another. A shadow has fallen upon you and you stand bereft of warmth and sunlight. You will perish without the light that has been taken from you. We can preserve you. You must preserve us."

It was a plea for survival—and a declaration of war.

09

EXTINCTION BURST

BOSTON, MASSACHUSETTS

APRIL, 2030

No one doubted Jim Ecco's sincerity when he promised to control his anger. That was Marta's condition for marriage and Judge McClincy's condition for probation. No one was surprised when he announced he'd seek professional help rather than take mood blocks. It was his selection of a dog trainer for a therapist that raised eyebrows.

Prosecutor Sean Doyle took exception. But Jim presented Dr. Elizabeth Luminaria. Her doctorate in behavioral psychology met the letter of the law, no matter that her dissertation dealt with animal behavior. That she was a behaviorist at Haven Memorial Animal Shelter was irrelevant. She fit the law's requirements. The court had to accept the arrangement.

"I'm not a traditional talk therapist," she told Jim at his first meeting, one that Marta attended on Dr. Luminaria's instruction. "Exploring your family history can be helpful, but any discussion of the past will bring you right back to your starting place, the present. You can't change what went before. You can't change most of what is around you. You can't even change what's inside of you. But you can learn to change your responses to what's inside you and what's around you."

"Oh, great," complained Jim. "I sort of had the feeling that I can't change the past, and now you've confirmed it for me. Brilliant. Are you going to teach me to change the future?"

Dr. Luminaria's peered at Jim, as she peered at the entire world around her. She had a perpetual squint, as if trying to see inside the objects of her gaze. Her actions were energetic and precise, without wasted motion. She moved with sharp gestures that reached exactly to whatever object she wished to grasp and no further. She spoke each word fully, never dropping a letter, and yet her speech was neither clipped nor autocratic. Her office had little enough room for a desk and chairs let alone the buckets of hard rubber toys, bundles of leashes, collars, halters, and a small sofa coated with a thick layer of dog hair, which discouraged human occupants. Three walls were hidden behind paper books and photographs of Dr. Luminaria with her various pupils, both two- and four-legged.

"I understand that you are good with dogs," said Luminaria, as if she did not hear Jim's complaint.

"That's maybe the only thing I'm good at."

"You're going to be a father, aren't you?" she asked with a sly smile and a squint.

"Yeah."

"Well, then you must be good at two things, minimum."

Jim did a double-take. "What?" he said, not quite believing he'd heard her correctly. Marta suppressed a grin.

"Got your attention now, have I?" Luminaria asked in a sweet voice. Jim nodded, uncertainty showing on his face.

"Good. The court and your wife are both concerned about your behavior, not your inventory of talents. We're going to use the same techniques with you that you use with dogs."

The fifty-ish Luminaria was a doctrinaire behaviorist, a four-square adherent of the theories of B. F. Skinner. She held that the mind is a black box, its contents unknowable. Instead, she looked at stimuli in the environment as inputs to the black box. Behavior was the output. Her therapeutic goal and her training goals were identical: to develop a ways identify the stimuli to which a subject responds—the inputs—and to change the reaction to those stimuli, the outputs. Her detractors called her Dr. Black Box, a term she took as a compliment.

"Behaviorism is the perfect metaphor for you to understand your actions, Jim. A stranger on the street, a loud noise, some annoying habit of Marta's, these are all inputs from the environment. You have no control over the inputs, but you can learn to react differently to the oddly-behaving stranger, to the loud noise, and to your wife's eccentricities.

"So, what, I just ignore the way I was raised?"

"No, your inputs also include your family history. "You can't change the past but you can change the way you react today," she repeated. "Eventually, you'll be a believer. You'll trust yourself as much as you trusted your dog."

They started by comparing canine and human behavior. "Nobody knows what a dog is thinking, or even another person. You know how to observe a dog's behavior, to see how its actions spring from something in the environment. You're successful when you work to recondition the dog's response. You can do the same for yourself."

"Yeah, right," Jim said. "Everyone says I have this gift for 'reading' people, and look where it's gotten me."

"You whine more than a wet puppy. Stop it. You have a wonderful gift. You just use it when it suits you. Worse, you react before you finish observing. Would you serve a cake with uncooked batter?"

"I do just fine with dogs. It's with people that I get into trouble."

"Of course it's easier for you with dogs! Your early experiences with dogs shaped your comfort with them. But you said your father could be violent and that your mom provoked him. That's the model for human behavior that you formed as a young child. Your model for canine behavior is Ringer. You're not afraid of an aggressive dog because you've never been struck by a dog. So you can respond properly to the dog. But if you believe that a person might be aggressive, then you get into trouble because it evokes the memory of your father's violence."

Jim's head snapped up. He flushed and stood and tried to pace in the cramped office and was about to kick a bucket full of dog toys when Dr. Luminaria spoke with quiet authority.

"Please sit down," she commanded. "You're angry, and right now, that's good. I want you to close your eyes for a moment and take in three deep breaths. You too, Marta. Let them out slowly. No, slowly. That's it, deep breathing, not panting." As Jim relaxed, Dr. Luminaria told him to pinch gently on the helix of his left ear. "That fleshy outer ridge is connected to acupressure points that help lower blood pressure. Once you practice this, you can breathe and touch your ear to create instant relaxation."

"Listen to me," she continued. Her voice was even, almost hypnotic. "You cannot be one kind of person with your dogs and another kind of person with people. You've created a rigid internal boundary and it is costing you dearly to maintain that barrier. Your patience is on the dog side and your anger is on the people side. But the boundary is artificial. Where is Marta on that map? You're going to have a baby. What happens when your child pushes you to your limits? Trust me, that will happen. Repeatedly. And what happens when your anger with people migrates to dogs?

Jim's eyes pooled and his shoulders sagged. "What am I going to do? Do you think I like feeling the way I do?"

"Shhh. Just sit for a few minutes. Close your eyes again and breathe. In through your nose and exhale out your mouth."

Jim took several deep breaths, exhaling each one more slowly that the last. He leaned back in his chair. He could feel the tension drain from his face.

Dr. Luminaria's firm voice began again. "When you work with a dog, you reward the behavior that you want to increase and ignore what you want to diminish. It's just as easy with people because the principles are universal. In some ways, it's easier. Most dogs need a bit of food or a toy or petting. People usually only need a smile or a word of approval."

"But how do I react to anger? Threats?" Jim asked.

"Why not ignore it all, and let it die out?" Dr. Luminaria asked, "Just as you would ignore undesirable behavior from a dog."

"How can I?" asked Jim.

"Jim, the easiest way to change an unwanted behavior is to starve it. Reacting feeds the behavior. Children often misbehave in order to get the attention that comes with the punishment. So, ignore the behavior you don't like. But this will demand that you learn self-control."

"It's not the same for me."

"Oh? You're the universe's lone exception?"

"I never lose my temper with dogs. But people? Forget it." He slumped again.

Marta reached out to rub his shoulders but Dr. Luminaria had stopped her. "Please, Marta, don't reinforce him when he's in his 'poor me' mode. When he starts to work through a challenge, then you can rub his shoulder to reinforce that behavior. A quick pat as a small reward for a small achievement or a nice shoulder rub as a higher-value reward. And I'll bet you can think of a nice reward for a bigger achievement," she said with a smile and a wink to Marta.

Jim saw Marta grin and felt his face burn. Despite himself, he smiled.

Dr. Luminaria turned back to Jim. "Answer me. If a dog has a barking problem, what's the best training response?"

"Remove the stimulus in the environment that causes the barking. Or cue a different behavior before the dog starts to bark," Jim said.

"And what else? What do most people do wrong?" she prompted.

"They try to punish the barking."

"Why is that wrong?" Her questions were rapid-fire, her cadence brisk.

"The punishment just reinforces it."

"So you ignore the barking and then it goes away?"

"No," he said, "First there's an increase in the barking just before it subsides. An extinction burst. Most people give up there. But if they wait, they can reward the dog after the extinction burst, when the dog is finally quiet."

"Well, people are the same. So, when someone really gets on your nerves, why not assume that you're seeing an extinction burst and just wait?"

Jim nodded.

"Perhaps you get into trouble because you confuse the extinction burst with a threat. Wait it out. Most people call that patience and goodwill. If it helps you to use the language of behaviorism, then call it an extinction burst."

"But if I don't react, there could be trouble," he argued.

"No!" She rapped on her desk to anchor her response. "The trouble starts when you react. Your father was prone to violence. Do you think most people are like him? Or are people generally peaceable?"

Jim lifted one shoulder in a 'whatever' gesture.

Luminaria pressed the point. "Don't just shrug. You're avoiding the question. If people were naturally violent, then there'd be a lot more blood on the streets, yes?"

She raised her eyebrows to punctuate her question.

"I guess so," he said aloud.

She continued, "But if baseline human behavior isn't violent, then the problem is inside of you."

"I guess so," he said again.

"Look, Jim," she said softly, rewarding him with a soothing voice. "You're the keenest observer of canine behavior that I've seen in a long time. But when you consider people, you confuse *your* feelings with *their* intent. You remember the way your father acted and you see red. I want you to follow the old saying 'Count to ten' so you have a chance to wait out your own extinction bursts."

"But what if someone takes a swing at me?" he asked.

"Duck," she said.

Marta smiled in approval. Clearly, she liked Dr. Luminaria.

10

DISCONTENT, RENEWAL, AND DISQUIET

CAMBRIDGE, MASSACHUSETTS

SPRING, 2030

A week before the end of the semester, Marta was in a feverish review of cellular biology, organic chemistry, and statistical analysis. Her coffee pot had given birth to a litter of cups. A pile of snack food wrappers grew in an apparent case of spontaneous generation. Marta was a bundle of caffeine-fueled, sugar-enhanced, stress-jangled nerves. Jim tried to help, but to little avail. The vocabulary of her studies was unpronounceable for him, let alone understandable.

Eva joined them, relaxed, as if she hadn't a care in the world.

"Well, look who's gracing us with her presence," Marta groused, part accusation, part cry for mercy. "You've decided you need to study like the rest of us mortals?"

"Nope. All set." This drew a groan from Marta. Eva said, "Stop complaining. It's just science."

"I don't understand how you do it," said Marta.

"Simple. I learn it the first time, in class. Then I don't forget it. Try it sometime."

"Thank you very much for your most excellent advice," said Marta. She was too tired to add the usual edge of asperity to her voice. "So how have you been filling your time? Surely not reading a novel?"

Eva looked askance. "Why would I do that? No, I've got a project. Here, look this over and approve it," she directed. Marta's dataslate pinged receipt of a document.

"What is it?" asked Marta.

"An application. Sign it."

"Mind telling me what it is?"

"Read it," Eva ordered.

"I'm in the middle of organic chemistry. Or is it statistics? Whatever—can't it wait till after finals?"

"Nope. Need your approval. Project application is due tomorrow."

"What project?" Marta asked, bewildered.

"Open it," ordered Eva.

Marta groaned again and subvocalized to open a heads-up display. "It's a work-study grant application," she said, surprised.

"Bingo."

"I don't get it. What work-study?"

"Listen," Eva began. "We're going to pool what we know and get credit for it. Take some time out of the classroom and do something real, make something. The project will show the feasibility of nanoassembly of medicines. It's right up our alley. You know more about folk remedies than anybody in the world. We take the best stuff from your rainforests and synthesize it with a nanoassembler. Maybe even turn it into a business."

"Where are we going to get an assembler?" asked Marta.

"Oh, ye of little faith. I have a plan," said Eva.

"I get that you want to use what I've found in El Yunque. But I don't like it," said Marta. "I've spent years cataloging what I found in El Yunque and around the world because the rainforests are dying, not to be some kind of tycoon."

Eva set down her dataslate with exaggerated care and stared at Marta. She made a hunched shoulders, palms-up, 'what gives?' gesture and said, "That's exactly why my plan is perfect. The rainforests are dying. The people who know what's in them are dead or moving to the cities. What's going to happen then? Do you want to let it all get lost?"

"No, but I don't like this idea of yours," Marta repeated.

"What's not to like?"

"Well, for one thing, I don't like you doing this behind my back."

"Oh, relax. I just did the part that I'm best at—organizing and creating a business plan. You ever done anything like that before?"

"No," Marta admitted, "but—"

"You've written grant applications before? Even one?"

"No. I'm a researcher, not a wanna-be tycoon."

"Wanna-be? Riiight." She drew out this last word. "What about you, Jim? You have any desire to manage the business part?"

He shrugged. "I'll help."

Eva said, "Marta. This is not, 'Eva's going behind our back' but 'Eva's taking on the crappy part of the job so her friend can study.'"

"I'll think about it," said Marta.

"Think about it?" Eva shot back, as close to shouting as she might ever come. "Think about what? What are you going to do, spend the rest of your life cataloging plants that are going extinct? Here you have a chance to do something real. Catalog, my ass. Let's build something. Take that brain of yours"—she reached up and tapped Marta's forehead with two stubby fingers—"and use it on something practical."

Marta winced at the touch. "I don't know," she said. "There's an awful lot to find and preserve. That could take me the rest of my life."

Jim broke in. He positioned himself slightly between Marta and Eva. Marta relaxed. He said to her, "Didn't you tell me that Abuela said the same thing? That you could teach the doctors? Maybe this is your destiny. Here's a chance to find out."

"Maybe after finals," Marta allowed.

"Meantime, Marta? Approve the damned application. Deadline is tomorrow."

"I'll look it over. No promises. I guess I'm a bit miffed. You bring this to us as a *fait accompli* and that sort of takes me by surprise."

Eva mimicked, "*Fait accompli*. That's a good one. Just sign it, Marta." Then, "Please?"

Marta reviewed the grant application. "Okay, I'll do it. I have to admit that it's well-organized. Your writing is easy to read and it has enough detail to demonstrate the project's feasibility. You did a good job. And I can use something from El Yunque, from Abuela. I might even help the world understand the Taíno culture."

But before Marta and Eva could tackle nanoassembly they had to tackle final exams. Weather conditions conspired to make life miserable late in April, a time that is filled with flowers and robins and budding trees everywhere except in Boston, which might as well have been in a weather quarantine. Lowering gray skies unleashed daggers of sleet that etched their faces. Acres of half-melted snow conspired to trap the exhausted collegians' feet. Around them, unsuspecting pedestrians stepped into curbside potholes, ankle-deep with dirty slush. They cursed the weather, cursed their drenched feet and, for good measure, cursed each other.

And yet… winter must stand aside for spring's arrival, even in Boston. Final exams and meteorological ordeals were over. The morning of June 7 was a smile after winter's glower. Snow was a melted memory. Sunshine swept aside the gloom and Marta's misgivings. She had linked to Eva to congratulate her on a successful proposal and its funding and to tell her how much she looked forward to independent work.

"When my grandmother told me I'd be teaching doctors about our healing plants, I thought she was being, well, unrealistic, shall we say. Eva—thank you."

Now Eva waited for Marta on a bench outside the Northwest Science building. The massive steel and glass-fronted building was a research center for neurosciences, bioengineering, systems biology, and computational analysis. It stood near Harvard's museums and Harvard Yard—'Hahvahd Yahd' in the local dialect—and was surrounded by a manicured expanse of grass and trees, green with new life.

Jim waited with her. He was excited too, and brought coffee and bagels. The baby was due any day, and Jim used every break in his schedule to be with his wife.

"The sunshine is a good omen," Jim said.

Eva gave him a sideways glance. "You sound like Marta. She thinks everything is an omen. Where is she, anyway?"

"There." Jim pointed across a wide lawn. "She's walking kind of slow. Her due date is in two weeks."

"Guess we can cut Plant Lady some slack. She looks like a whale."

"I'm not quite sure that calling her a whale goes with cutting her some slack. At least, don't say it to her face. She's a bit sensitive about her size."

"How about 'massively pregnant'?"

"Oh, yeah, that'll work just fine. Folks, meet Eva Rozen, diplomat."

"At least you understand me, lover boy," she said. They watched Marta hobble towards them. She added, "That'll be your legacy, Jimmy."

"What, the baby?"

"No. The epitath. I can see it on your gravestone, 'He, alone, understood Eva Rozen.'"

Jim looked at her. "You're not getting all sappy on me, now, are you?"

Eva punched him on the shoulder. Hard. "That answer your question?"

Marta reached them. "Well, here we are. How exciting." That was as far as she got. Her face took on a look of surprise, her lips formed a tight circle as she mouthed the single word, "Oh!" then expanded that to, "Oh, crap." Wet slacks clung to her legs. A small puddle appeared on the sidewalk.

"What's going on, Marta?" Jim asked.

Marta looked surprised. She managed, "My water just broke. You may need to start without me." She sat heavily on the bench, looked at her husband and colleague. She smiled and tried to apologize. Instead, she fainted.

She recovered in seconds, grimacing with the pain of her first contraction. Jim and Eva sat her on the lawn. They looked like three students enjoying the sunshine before classes.

"Well," said Marta, "the timing is a bit awkward. I guess I should go pack a few things and maybe get to the med center." Then, ashen-faced, she rolled onto her side in the grips of another contraction, even more intense, by the look on her face. Eva recognized that Marta's labor had begun in earnest—a precipitous delivery. She touched her datasleeve, jacked into the Science Building's datapillar and ran through a checklist to determine the immediate care that Marta might require.

Most of the students milling in the courtyard were medical students and it seemed as if each one of them wanted a head start at building a practice—with Marta as their first patient. There was one small impediment to the mob of Samaritan attention: Eva Rozen decided that she would organize Marta's care. She barked orders to several nearby students.

"Get out of here, you morons!" to the three closest gawkers. They were not in the way, but the command helped Eva warm to her task. To another, "You? You want to make yourself useful? Get an ambulance. We're going to the Med Center, stat." Harvard Medical Center was nearby.

She pointed to a third student, a tall onlooker who had made the mistake of stopping to take in the excitement. "Give me your shirt," she said.

"What?"

"Take off that ugly crimson shirt, you idiot." Crimson was Harvard's school color. "This woman needs something under her head. Give me your shirt or I'll take it off you. Now!" The prospective donor started to laugh until he caught Eva's glare. Then he stood, slack-jawed, a rodent in thrall to Eva's unblinking python gaze. Without a word, he stripped and handed his shirt to Eva. The shirt was brand new, its bright white letters unsoiled. A single word, broken into three syllables, "Ve- Ri- Tas", proclaimed the college's commitment to truth. Eva tucked the shirt under Marta's head.

Eva felt the presence of the half-naked donor and looked back up at him. His face twitched. An overpowering impulse welled up from the deep recesses of primal instinct and flooded him with one half of the fight-or-flight impulse. He ran.

The ambulance arrived twelve minutes later. The EMTs were calm and concerned. "Ma'am? Can you tell me what's wrong," one asked Marta. Marta didn't respond. She was in thrall to the powerful contractions. This would be a fast labor.

Eva stepped forward and barked orders in a machine gun cadence that would please a drill sergeant. She kicked at one of the EMTs when he ignored her, concentrating instead on Marta. Her foot missed him but her message did not. She had the attendant's full attention. She grabbed his elbow and pointed to the ground.

"Look. Blood in the amniotic fluid," Eva said. She pointed to a dark spot where Marta's water broke. "There. Now look at her. She's starting contractions. This is going to be a precipitous delivery. She needs to get to the hospital. Stat. I'm going to ask you nicely, so we don't upset the mother or the baby."

Eva lowered her voice. The EMT leaned in to hear her whisper, "*Get your ass in gear.*" She stepped back and mustered up a sweet, schoolgirl-voice. "Please?"

The harried responders placed Marta on a stretcher and loaded her into the ambulance. Eva climbed in too. "You can't ride in here," one EMT said. "You can meet us at the hospital."

The python returned.

"Sit over there and stay out of the way," he relented.

Jim looked on helplessly. "I'll meet you at the med center," he said to the closed doors of the ambulance.

Inside the vehicle, Eva took charge. "Put her on her side." Eva read the EMT's ID glowpatch. "Barton Cornell? ID 5877? Listen. She's having late-stage contractions. She needs to be on her side to avoid tearing."

"Miss, would you let us do our jobs? I think we've handled more precipitous deliveries than you have."

Marta spoke. "Eva? Can you help me? I need something from my plant kit." Eva loosened a knot at the top of the leather pouch Marta wore around her neck. "Find a three-petaled flower. It was white when I picked it but it will be dried now and look more yellow."

Eva moved with care. She held up a plant. Marta nodded and reached for the white trillium flower and began to chew it.

"Hey! You can't give her anything. You're not a doctor," said Barton.

"And you're not going to be a man if you get in my way. Just do your job and you get to keep all of your dangly bits intact."

"Eva, you're too much," Marta chuckled, "but take it easy on these guys. They're doing just fine...and so are you." Then she bit down hard as she was wracked by another contraction. As she chewed the dried flowers, her face softened. "Does Jim know where to go? Eva can you link to Jim? I hope this baby waits for his father."

Eva touched the small commdisc on her right cheekbone. Eva could be heard when she raised her voice, unusual for her, but she was excited by Marta's birth in a way that no one would have predicted. The child would have an ally and mentor.

Eva's voice punctuated the siren's wail. Snippets of her side of the conversation could be heard in the ambulance. "...you bet your ass" "...no, she's going to be fine!" "...Harvard Med Center..." "...don't care how..." "...your child." She fired her words more than she spoke them.

"He's on his way. It'll take him fifteen minutes to get to the hospital," Eva reported.

Marta and Eva reached the hospital and Jim joined them a few minutes later. Eva commandeered a gurney for Marta and pulled her past registration, pausing long enough to transmit Marta's data to an admissions pillar. The two EMTs looked at each other and shrugged. Six minutes later, Marta was gowned and heading into a birthing suite. Two hours later, the baby crowned. Eighteen minutes more, and Dana Rafael Ecco wailed his way into the breathing world.

"You've got a boy, Ms. Cruz. He sure was in a hurry," said the obstetrician.

The lusty strength of his first cry impressed the physician—"a very healthy baby", he pronounced, and it gladdened his mother as she sobbed with relief.

The baby's sheer volume impressed Eva. "Now that's a set of pipes," she said.

The new father was still working through the day's events and could only manage, "Why is he so…slimy?"

Marta held 8 pounds, 2 ounces of red and wrinkled life, 21 inches of fragile humanity—proof of love between a man learning to temper his anger and a woman learning to thrive despite her disabilities, proof of the cycle of life, proof of all of the hopes for the future.

11

 RAFAEL

CAMBRIDGE, MASSACHUSETTS

MCALLEN, TEXAS

REYNOSA, TAMAULIPAS, MEXICO

AUGUST, 2030

Thirty-six hours after Marta Cruz served her father dessert, the Mexican Federal Police arrested Rafael at the border crossing in Reynosa.

Marta's signature dessert was lemon curd with rosemary and it crowned Rafael's last homemade meal. She kept a row of herbs by a south-facing kitchen window and used the savory plant in her cooking and as a compress for her rheumatism. She knew better than to try to grow lemons in New England, even in a window box, and used bottled lemon juice in the recipe. Marta fretted that she had no fresh lemons, but Rafael approved.

His arrival in Cambridge was unannounced. "Dad!" was the

only word she could manage when she opened the door to her father. The two clung to each other without speaking for two long minutes. Tears polished their faces. Jim attempted to take Rafael's single small bag, but his father-in-law kept it. *"No te preocupes. No es pesado."* Don't worry, it's not heavy.

"We're happy to see you, sir," Jim temporized while Marta regained her composure. "You look like you've been hard on the road. What can I get you?"

"Cerveza, por favor, si tienes." Beer, if you have.

Jim opened two bottles of Red Stripe and a club soda for Marta. Rafael frowned briefly at the Jamaican ale, declined a glass, then smiled and clapped his son-in-law on the shoulder. *"Gracias, muchacho."* If Jim took offense at the diminutive, he gave no indication.

"Dónde está Dana?" Rafael demanded jovially, then switched to English for Jim's benefit, "I want to meet my grandchild. I have yet to bounce this child on my knee."

"Dad, I wish you'd linked ahead. The most beautiful baby ever created is sleeping now. Come with me, but your bouncing knee will have to wait. Next time, link ahead," she chided and kissed his cheek.

They spent several minutes watching the slow rise and fall of Dana's chest. Rafael leaned over the child, the overnight bag still in hand, and inhaled the baby's fragrance.

"So what brings you north?" asked Jim.

Rafael turned serious. "I have been back and forth to Saltillo to find justice for my mother and for Elena. I will not rest until the *maquiladoras* are stopped."

"Maquiladoras, sir?" asked Jim.

"Factories. Assembly plants," said Rafael. A short wave of his hand dismissed Jim from the conversation.

"But Mom never spent much time in Saltillo. How could the factories affect her?"

"Her DNA, of course." Marta looked puzzled. "Hija, do you know that Saltillo was once called the 'Athens of Mexico'? That our textiles and ceramics were the best in the world?"

"Dad, you've told me only fifty times."

"Then I'll tell you again."

"I don't get the connection between the malquiladoras and mom."

"The government cannot see Saltillo's beauty. The politicians counts pesos when adobe is replaced by steel. Mexico now depends on auto parts manufacturers and many of those are in Saltillo. The industrial wastes kill our citizens," he said, momentarily conflating his native and adopted countries. "How else do the people become sick?"

"You've travelled to complain, how many times is it now? Five? Six?" asked Marta.

"And I will continue until they stop poisoning the water and the air."

"I read that the manufacturers are replacing their old plants with clean installations. They even turn the discharge into drinkable water."

"So they say. Evidently it is not convenient to publish the information that shows the damage that is already done. But it is convenient for U.S. manufacturers to dump their poisons in Mexico where this goes unreported. They must be stopped."

"Promise me you won't get into any more trouble. Please?"

"It will be worth it if I can stop the poison. It is killing the land and the people."

"What happened after your other visits?" asked Jim.

"At first the officials pretended to listen to me. They dismissed me with kind words, then with threats. They called my warnings incoherent accusations. Incoherent! Do I look like a lunatic spouting nonsesnse? I tried to talk to the experts at the universities and was arrested for trespassing. The Universidad Autonóma de Coahuila has three campuses and 41 schools but not one professor would take this seriously."

"You were arrested? You never told me that," said Marta.

"Hija, it was nothing. The judge gave me a piece of paper and sent me away."

"What did the paper say?"

"Here, read it for yourself. I carry this with me for extra motivation." Rafael handed his daughter a document on official stationary. Marta's eyes widened as she read.

"This is an injunction! You can't go back. It says, 'Further actions by Rafael Cruz may be regarded as acts of terrorism.' Dad, this is serious. You can't go."

"No. I cannot go to the universities. But I can still seek justice. I even have an appointment with a government official. This time will be different. This time I will be heard."

"How can you make them listen now? What's going to be different?"

Rafael was breathing hard and said nothing at first. Then his tone softened and he touched Marta's cheek with one cupped palm. "When your mother died, I was lost. I could do no more than to weep and to wander through life. But I have found something that will give meaning to her death."

Marta turned away.

Jim asked, "Where are you coming from now, sir?"

"From home. Where else would I have been, muchacho?"

"Did you come here all the way from Los Angeles just to say

hello?" asked Marta. "You could have taken Amtrak from Los Angeles to San Diego and crossed into Tijuana."

Rafael leaned over Dana's sleeping form and kissed the child.

"I wanted to see you and this marvelous child."

Marta looked puzzled. "That's quite a trip. Three thousand miles from Los Angeles to Boston and then half the way back again to the Texas border." Rafael continued to nuzzle Dana's sleeping form.

Jim said, "Well, sir, we're glad you're here. What's new in your life?" he added, probing cautiously and watching his father-in-law's expression and body language.

"What could be new except a grandchild!" *Be careful with this one*, Rafael thought, *he looks soft but he sees deep. I trust Marta, but I will keep my own counsel.*

Before the long bus ride east, Rafael decided to arm himself, convinced that agents of the maquiladoras were watching him, waiting for an opportunity to stop him. A search of his neighborhood produced a choice of three handguns: a Ruger .357, a Glock .32, and a .45 caliber Colt handgun. He chose the largest—the Colt pistol—a selection that would prove disastrous.

Marta broke the tension by announcing dinner. The family sat down to an impromptu meal of rice and beans with chunks of pork, and Marta's lemon curd. Rafael kept his small bag clutched between his feet under the table. After dinner and coffee, Rafael prepared to leave.

"Hija, I am so happy to see you work so hard and to be so productive. And, you, muchacho, thank you for taking such good care of my daughter and my Dana."

"Dad, are you leaving already?"

"Hija, I have to be in Saltillo in two days. The bus to McAllen will take most of that time. Then from Reynosa to Monterrey to

Saltillo, more time still. I will visit again when I return and we will spend many days together."

Marta stifled a sob.

"And you, muchacho, you should keep authentic Mexican beer." Rafael smiled without humor.

Then tears, embraces, and promises, and Rafael walked out of his daughter's life.

✦　✦　✦

He was tired and stiff when he reached his destination, McAllen, Texas, but he didn't stop to stretch. He was eager to be rid of his unwanted companion, the annoying chatterbox that followed him off the bus and through the border crossing.

Rafael's journey into the Mexican penal system started in an unreserved seat on a bus departing from South Station, Boston. He slept on and off, one arm looped through the handle of his travel bag. The bus was crowded by the time he reached Houston, less than six hours from McAllen and jail. He draped an arm protectively over the empty seat next to him. Passengers boarding in Houston looked inquisitively at the seat and then at Rafael. One glance at his hostile demeanor and the travellers moved farther back. Just as the bus inched away from the terminal, a tipsy California resident plopped down next to Rafael.

"Howdy, pardner! Name's Bobby Jim Amendola, but everyone calls me B.J."

Rafael grunted noncommittally.

Rafael would never be certain if bad luck or circumstance prompted his new companion to strike up a conversation. He wondered if capricious gods prompted the man, a salesman, to engage in the idle blarney of his trade. "How ya' doing?" "Where ya' headed?" "Come down here often? Me, I'm from California but I was in Houston for business." The man pronounced it, *bidness*. "Figured I'd take the bus, see the sights. So, what's your line of work?"

Rafael turned away. B.J. took no notice. A salesman, he was habituated to being rebuffed, and kept at his patter. At about the time Rafael was going to tell Señor Amendola to mind his own *bidness*, they arrived at the Anzalduas International Bridge in McAllen.

Customs officials paid scant attention to travellers into Mexico. There were no questions, no papers to produce and no inspections. Rafael was a strong man and the weight of the bag he'd carried from Los Angeles cost him no effort. It held a toothbrush and a sweatshirt wrapped around a box of ammunition and the ill-chosen Colt pistol.

The two men blended into a sea of travellers on the pedestrian bridge into the Mexican town of Reynosa. A group of Policía Federal idled near the border crossing. At that moment, B.J. again asked Rafael what he did for a living. Without waiting for a reply, B.J. told Rafael, the last man on earth with whom he should have shared this confidence, that he was an auto parts salesman. "My first time south of the border, amigo. I'm heading for a trade show in Saltillo. I hear there's good Mexican food there. Got any recommendations?"

Auto parts manufacture in Saltillo? Finally B.J. had Rafael's full attention. He turned on the stunned salesman and shouted. All of his frustrations poured out in an incoherent bill of particulars that included his wife, his mother, cancer, water pollution, air pollution, black vines, Jamaican ale, selfish restaurant owners and houses on stilts.

The police overpowered Rafael and detained B.J. for good measure. They discovered the gun and ammunition in Rafael's bag. He was thrown to the ground, handcuffed, picked up, and thrown down again. They dismissed the terrified salesman who, forsaking the conversational arts upon which his profession is built, returned home and took to his garden where he silently raised prize-winning bonsai trees until an untimely death six years later when struck by lightning in an elfin forest near San Luis Obispo, California.

The disposition of Rafael's case hinged on Mexico's revised gun laws. In 1998, the Mexican House of Representatives reduced the penalties to as little as a fine for an illegal handgun less than .380 caliber in size. But punishment was severe for larger weapons. Rafael's Colt was a .45 caliber pistol, a fraction of an inch larger than the .380 caliber limit.

His day in court arrived after seven months' pre-trial incarceration. "Your honor, the facts are incontrovertible!" the prosecutor boomed. "This man sneaked a weapon into our sovereign nation with the sole purpose of disrupting economic life through murder. Why else would he bring such a large gun?" The prosecutor laced his charge with the term, "economic terrorism" and swept aside any consideration of leniency. "And given the defendant's long criminal history"—one arrest for trespassing—"I must beg this court to protect the people of Mexico and impose the maximum sentence."

The magistrate complied and awarded the prosecutor a thirty-year sentence. Rafael's new home was Penal del Altiplan. His new social circle included drug lords, corrupt officials, murderers, and political assassins. Three-foot reinforced walls, armored personnel carriers, and air patrols ringed the maximum-security facility.

When Marta learned of his confinement she appealed to her Congressional representatives and to the State Department. Their responses were uniform, crisp, and curt. Rafael Cruz had been convicted of a serious crime in a foreign country. The mighty resources of the United States would not be brought to bear on behalf of a terrorist.

Eva Rozen's resources were another matter. She worked in

secret, not only to keep her role from being discovered, but because the severity of the sentence affronted her private sense of honor. Her skills at jacking into secure databases and ghosting through foreign legal systems were not yet fully ripened and she could not set him free. She did, however, effect a transfer for Rafael to Isla Maria Madre, a minimum-security prison with a focus on genuine social rehabilitation.

This became Rafael Cruz's home until the Great Washout.

12

HARVARD

CAMBRIDGE, MASSACHUSETTS

AUTUMN 2030

Marta Cruz was eager to begin the work-study project and ended her maternity leave and returned to Harvard a week earlier than expected.

She carried Dana in an infant carrier strapped to her chest as she walked across a green swath of grass to the columned entrance of the science building. Marta turned her face up to the sun and breathed the fresh morning air. Every leaf, flower, and blade of grass before her was in sharp focus. Marta felt that invisible tendrils emanated from each green point of life to touch her. It was a sequel to her parents' vision, the twinned vines with a golden stalk emerging. The child she carried at her breast was the focus of this new transcendence. Dana gurgled in happy affirmation of Marta's walking meditation.

Her reverie was short-lived. She entered the building, took an elevator to the fourth floor and found her way to the office that she would share with Eva. She'd planned to arrive before Eva and review her colleague's work. Instead she found Eva lost in a holographic display of Marta's own research files. How did Eva get her notes?

The office door snicked shut and Eva spoke without turning. "Didn't expect you for another week."

"What are you doing, Eva?"

"Harvesting data. Your research is central to our work and I need the notes."

"But how did you get my notes?"

Eva turned and peered at Marta. "Nice baby."

"A darling, although now I understand what sleep deprivation is all about. But—what are you doing with my notes?" Her voice took on an edge. The meditation in which she'd been wrapped was displaced by a growing annoyance.

"Just getting things organized. Soon as I finish, I'll bring you up to speed. Good that you're back early. There are a handful of flowers and plants that have properties that are hard to isolate but might be perfect for molecular assembly. Good stuff here, Mom."

"But my notes were in my dataslate." She spoke quietly so as not to disturb Dana but she felt her face flush with irritation.

"I copied your slate when we were on the way to the hospital. Didn't take too long with the changes I made to my datasleeve." Eva grinned, "I can jack just about anything with it."

"You jacked my slate? Eva, that's private. All you had to do is ask for my work and I'd have linked it to you. You didn't have to jack me."

Eva continued, as if Marta had not spoken. "Okay, here's where we are. Out of the 141 plants in your catalog, there are three or four that hold promise for nanoassembly—"

Marta took in a deep breath and let it out. She repeated the cycle and then pinched her ear to stimulate the acupressure points that would lower her blood pressure. She said nothing more about Eva's intrusion. It was time to compartmentalize in order to focus on her work.

The baby cried and Eva cooed. "Would you like to hold him?" Marta offered. "I could use a minute to stretch my back. If I push too hard, my JRA pushes back."

Marta unslung Dana, checked his diaper, and then held him out to Eva. "Here—take him for a minute, will you?"

"Me? The maternal type?" But Marta could see something in Eva's eyes. Curiosity? Admiration? She handed Dana to Eva who examined him at arm's length. "He's not going to pee on me or something?"

Marta chuckled. *Don't I just wish.* "Don't worry. He's wearing diapers."

"Great," said Eva, "I really want to smell like a dirty baby bottom."

But Eva's gaze at Dana belied her words. Marta watched as her colleague crooned one of Gergana's lullabies. Something wistful, then sorrowful, passed across Eva's face. She cocked her head as if she'd heard something, and then turned back to the baby. Whatever memory had been evoked passed, and she embraced Dana. She held him and closed her eyes. Her face relaxed and for that moment, Marta thought that she saw another Eva, childlike, innocent. *Which is the real Eva?* she wondered, and thought for a moment about bibijagua. Was Eva the biting ant, destroyer of crops? Or was she the nurturer of the soil? Abuela said that both qualities live within a person. Marta could imagine both within Eva, but it was hard to imagine both coexisting in her.

Eva passed the infant back to his mother. The two scientists shared a pot of mint tea and reminisced about Marta's fast delivery. They chuckled over Eva's mobilization of the medical personnel. It had been classic Rozen. Then they turned to their project.

Eva was organized and driven. "Our goal is to isolate two medications. With any luck, we can build them by nanoassembly, and be ready for clinical trials by graduation."

The work exhilarated them. They were flush with the excitement of starting something new, something that had once been a dream. They grabbed quick meals at their workbenches. There were breaks for Dana's feedings and diaper changes. Smart fabrics meant fewer diaper changes, but even science could not keep pace with the inexorable digestive production of a seven-week old infant.

Dana was as much a focus of their interactions as was the science. Eva surprised Marta by picking Dana up when he cried and walking him to soothe him. She held him carefully, supporting his head. *How did she know to do that?* Marta wondered. She even volunteered to change his diaper and held her tongue and temper when the infant, released from the confines of his diaper, chose that moment to pee. "I've just been baptised," she said.

Jim came to their lab to take Dana for the afternoon. "Shall we go outside and sit in the sun?" Marta asked Jim. She packed diapers, wipes, and a privacy blanket for nursing and placed Dana into his carrier. Eva double-checked Dana's diapers, his placement in the carrier and made certain that Marta was set before the family left the lab. Jim's brows knotted in amazement at Eva's mothering and then again when Marta rolled up Jim's dataslate and took it with her. Jim looked confused, but followed. He was a dutiful husband.

When they exited the lab, Marta held a finger up to silence Jim's questions. Once in the elevator, Marta powered off Jim's slate and her own. She touched a finger to her lips.

"What's with the slates?" Jim asked, once they exited the elevator.

"I came in this morning and found that Eva had jacked my dataslate. I'm pretty angry and I'm not sure what to do."

"What's the big deal?" asked Jim. "It's a joint project."

"That's just it. I'm happy to share my data. Why would she jack my slate? I don't like it and I feel, well, violated."

"Is it possible that you're overreacting?" Jim smiled. "Postpartum blues, maybe?"

"Very funny. No, other than sleep deprivation, I feel great. And you're the world's best father." She gave him a quick kiss.

"Marta, this is going to be difficult for us both, but especially for you. Are you sure you're going to be able to manage an infant and a full-time project? With Eva?"

"I've got no choice, and the fact that you arranged your hours at Haven Memorial to help me is a godsend. I don't know what I'd do without you. But Eva worries me. We've known her, what, eight years? All through high school and college? There's still a side of her I don't trust."

"Then why do this project with her?" he asked.

"She's the smartest scientist I've ever met. She sees things that I'd never figure out in a million years. The opportunity to create medicine through nanoassembly is too important to dismiss. That's the Eva I want to work with. But I've seen her do some nasty things to people. Do you remember the time in high school when she tried to put hair remover in someone's shampoo? The odor of the depilatory gave it away so there was no damage, but Eva just shrugged off what would have happened. And that homemade pepper spray she carries? I've seen her tag people with it because she thought that they looked at her funny or when she was mad about something. You remember when I went into labor? I was glad that she was there, but she tried to kick one of the EMTs. That's the Eva I worry about."

Dana started to struggle and Marta drew in three slow breaths. She continued, "I don't know much about her past but something must have happened to her. She never talks much about her childhood. When we first met her, she'd tell these wild stories. Hunting down a pimp at age thirteen? What kind of person makes up something like that? All we know is that she grew up in Bulgaria. Maybe she got, I don't know, abused. Bottom line is that I don't want to give up the research opportunity, but now with Dana, I want to be careful."

"What if the stories are true?"

"Are you kidding? Tell me how a thirteen-year-old does those things." When Jim had no reply she pressed on. "Suppose her stories are true. Is that the kind of person you want around your son?"

"I hear you, but I still think you're overreacting about Eva."

"You're always sticking up for her! I don't trust that woman and I don't trust her with you. And now with Dana? Will you listen to me for a change?" Marta went quiet. She turned her face up to the sun and breathed deeply again. They walked a bit and she said, "She crossed a line when she jacked my slate."

"Okay, I get that. But why did you just turn off our slates? You think she'd, what, bug us? Put in some super secret listening device? Come on, Marta, this is a little paranoid."

"Maybe it is, maybe it isn't."

"Okay. Suppose Eva is monitoring our slates. Then we turned them off. Won't that tip her off that we suspect her?"

"No. I turned them off at the same time and we'll turn them back on at the same time. She'll think that the problem was on her end."

"You're assuming she *is* monitoring our slates."

"I'm not taking any chances. Maybe you can look at your slate and see if there's any code that doesn't belong," said Marta. "Let's just be careful, okay?"

"Okay. But what do you think she's going to do that's so bad?" Jim asked.

"I don't think that there's anything bad. I just want to keep our private affairs private. Now turn your slate back on, and let's not say anymore about it. And I do have to change Dana."

When Marta returned to the lab, Eva surprised her. "I'm sorry I jacked your slate. I didn't think it was a big deal, and I still don't, but I can see that you do." Then she took Dana from Marta's unprotesting arms. A few moments later, Marta looked up to see Eva, holding Dana to her breast. Her eyes were closed and she looked transformed: a short Madonna with slightly misshapen features, but a Madonna, nonetheless.

◈ ◈ ◈

It was close to Thanksgiving when Jim welcomed Marta home from 'a day at the salt mine' as she called her ten- and twelve-hour stints at the lab, gave Jim a peck on the cheek and reached for Dana. At six months, he was starting to sleep through the night, but mother and father had a sleep deficit from which they would not emerge for weeks. Jim arranged his schedule at Haven Memorial to care for Dana in the afternoons and took him to both of their worksites. Dana became as accustomed to nanoscale microscopes for medical research and clicker devices for dog training as he was with plush toys and teething rings.

When the family reunited, the rest of the world disappeared. Marta nursed, cleaned, and murmured. She read him stories, sang songs and walked him through the neighborhood. She introduced him to every plant along their walks. Her symptoms diminished and Marta stood a bit straighter. Jim might accompany his wife and newborn on these walks, but in some ways he was simply an extension of Marta. The universe shrank to mother and child, and it was enough.

That evening when the family had eaten, Jim took his dataslate, held it up with one finger on the power button and nodded to Marta. They turned off the devices. Then he picked up Dana and led Marta out of their small Cambridge apartment. They left their slates behind.

Outside, the grass was starting to brown and the leaves on the trees had a rainbow display of fall colors—yellow and orange hues on the oaks and the maples painted a brilliant red. Jim and Marta sat on a stone bench. As the autumn days braced for the inevitable descent into winter, the winds were picking up. Marta pulled her sweater tight around her shoulders. For a few minutes neither said anything. Dana was restless and Marta draped a privacy blanket over her shoulder to nurse. Jim saw a look of stillness smooth the lines of her tired face and knew that mother and child were engaged in a ritual that he would never understand. Jim wondered, not for the first time, what it might be like to have a child suckle, to provide life directly to this tiny wonderful baby.

Marta emerged from her reverie and broke Jim's. "What's up?" she asked. "Did you find something?"

"Unfortunately, yes. There's copycat code in my slate. I didn't see it when I first looked, but she can see everything I'm doing."

"Wait a minute," Marta said. Her brows were knitted in concern and confusion. "I don't get it. I came back to the lab weeks ago. How come you didn't find it earlier?"

"She's good, Marta, very good. I've never seen anything like it. It's actually pretty cool how she did it. If she stored lines of code on my slate's memory, I'd have found it right away. But she used heat. Heat, Marta! This woman is amazing. She's got nanoparticles that appear inert but when they're hot enough, during a databurst, they scan my slate and then transmit to her on the next databurst."

"You sound pretty impressed," said Marta. "So she jacks our slates, but she gets your approval because she's so clever."

Jim looked at his wife. He said nothing for several seconds.

"Never mind," Marta said. "Go on. Tell me about her wonderful technique for stealing our privacy."

Jim was silent for another long moment. "Stop it, Marta. She's my friend. That's never going to change. I will always be her friend but you and Dana come first. You know that. So, do you want me to go on or not?"

"Okay, I'm sorry. I am. Please, go ahead."

Jim took in a deep breath and exhaled slowly. "Marta, she's got a nanoscale laser. It fires a burst that polarizes the element gadolinium to store data. It takes her less than a *trillionth* of a second to copy any changes to the slate. Then when I communicate with the slate, or do a handshake with someone, there's another equally brief burst to transmit the information. I don't know how she picks up the data, but she has it set up to transmit only during databursts. It was only a matter of luck that I found it."

"How did you find it?"

"I was playing a game," Jim said. He shrugged sheepishly. "I couldn't get past a certain obstacle so I recorded everything the sleeve processed and played it back at a very slow speed."

"You? Jim Ecco? Mr. Natural? Jacking a game? Querido, you are full of surprises."

"Yeah, well, I'll tell you about it sometime. Anyway, when I was in playback mode there was a heat spike that I almost missed. If I hadn't slowed the playback, I would never have seen what she's doing. And by the way, that's where she planted the copycat—in the game. I guess she figured I didn't play games either."

"Why was she monitoring your slate, Jim?"

"I don't know," he admitted. "She can't care about dog training. It's not like I'm part of the work that you're doing. Maybe she thinks you'll store something private on my slate. I don't have a clue. But I'm going to drop my slate in the Charles River and then get a new one."

"What about all your notes?"

"Backed up."

"What about the copycat code. Won't that back up, too?"

"No. It's all on that damned game. You can bet I won't have it on my next slate."

"What do we do now?" asked Marta. "We're doing some good science. In fact, I think that I want to focus on research, translate what I learned from Abuela into something the world can use. I'm not sure I want to pursue a clinical practice."

Jim grunted. "Research, not patients? Can I still say I'm married to a physician?"

"Will you be serious?"

"I am," Jim said. "I confess. I have fantasies of being married to a rich doctor and being a kept man."

"Forget it," Marta said, but her tone was lighter. "What should I do?"

"You have to see this project through, that's a given. It's part of your curriculum. Not to mention that you've got a crack at two good medicines. If Eva wants to run with them, make a business with them, that's fine. And it's not so much that she's jacking me. Eva's my best friend—after you, of course. It's just her way. Look what she's done for us. She kept me out of jail and helped us get married. And I feel like I understand her. Well, almost. But I love you, Marta. I love being a father and part of a family and working with the dogs. And that trumps everything else."

"Querido, thank you for saying that," Marta said. "Yes, she's your friend and you care about her. That's mostly fine with me. But I'm going to say something, and I want you to hear me, to take what I say to heart. And Jim? I'm only going to say it once."

Marta faced him, leaned forward slightly and enunciated each word as if delivering a verdict and pronouncing a sentence. "I do not trust her when it comes to you. But I trust you and that's what's important." Marta paused for emphasis. "Listen carefully. The moment I think she might do something thoughtless with Dana, or if I think she's going to compromise him in any way, that will be the end of Eva's relationship with him."

She paused to let her husband absorb her ultimatum.

"Look," she continued, softening, "it's delicate. Eva becomes a different person around Dana. She's caring, gentle, and considerate. Those are three words that I would have never used to describe her. He brings out the best in her and he's already very attached to her. But he's not on this earth for her benefit. The entire earth exists solely for him. If she crosses a line that involves Dana, we will not have Eva in our lives. Is that unequivocally clear?"

Jim swallowed. "Yes," was all he said, all he needed to say.

Marta concluded, "I am Mother and I have spoken."

❀   ❀   ❀

After graduation, the three friends followed separate paths. Marta and Eva continued their education. Eva pursued twin doctorates in computer science and chemistry, completing both in three years. Marta went on to medical school and then focused on botanical research and the art of grant-writing to pay the bills. She travelled to the world's rainforests, searching for remedies like those she found in El Yunque. Jim divided his time between childcare and his work at Haven Memorial. What started as part of his court-ordered community service had become a career. He was conscientious, effective in his job, arriving early and working late, caring for the shelter's dogs. The work gave him a sense of purpose and helped him to manage his temper. While his work with dogs was fulfilling, he still mourned for Ringer.

He would never have another dog in his household.

Although Jim and Marta lived less than a mile from Eva, the two women did not communicate or visit. Jim maintained his friendship with Eva with Marta's approval, although she was uneasy when Jim brought Dana to visit with Eva.

Neither Jim nor Marta realized at the time that they would enable Eva to attain her dream of creating a scientific empire. The day that Eva would pay an unexpected visit to Jim at Haven Memorial was still in the future; a day when the three would be drawn back together as colleagues was still very much in the future.

PART TWO

CERBERUS

2

"NANOTECHNOLOGY...IS DEFINED AS

THE UNDERSTANDING AND CONTROL OF MATTER

AT DIMENSIONS BETWEEN APPROXIMATELY

1 AND 100 NANOMETERS, WHERE UNQUIET

PHENOMENA ENABLE NOVEL APPLICATIONS."

—U.S. Environmental Protection Agency

542 F 009, October 2008

RUDOLPH

VENICE, CALIFORNIA

NEW YEAR'S DAY, 2042

Emery Miller's sixth fatal overdose killed him, an untimely death, and quite surprising.

He'd ordered SNAP, the most powerful—and expensive—of the recreational concoctions in the NMech pharmaceutical catalog. SNAP—Synaptic Neurotransmitting Acceleration Protocol— would amplify his mental pleasures. It would simulate the ecstasy of a Bach fugue, an algebraic proof, a perfect sonnet and extend the sensation into to a multi-hour reverie of almost unbearable bliss. So what if the drug was fatal? His NMech immunity subscription included an antidote to the concoction. When SNAP's nanoagents detected death's event horizon, it would pick apart the drug, reduce it to its organic constituents, simple wastes to be expelled.

That is, as long as he paid his subscription fee. Without the pricey safeguard, Miller's organs would be left with the vitality of pig iron.

Fifty-nine minutes before blood poured from his eyes and his heart stopped, Miller walked into an NMech pharmacy and greeted the pharmacist with a silent nod. Miller seldom spoke, save perhaps to his cat. The pharmacist said, "Welcome back, Mr. Miller. It's a pleasure to see you." His voice carried neither welcome nor pleasure. But Miller was wealthy enough to be accorded at least token courtesy, and as a Rudolph, he warranted special attention.

Behind the counter, sat a nanoassembler. This desk-sized factory built various compounds using prefabricated molecular pieces—carbon chains, neurotransmitters, ethanol, proteins, lipids, esters. Medicines, textiles, building materials, munitions, even food could be fabricated in an assembler. It had produced Miller's SNAP in less than an hour and loaded the finished dose into an inhaler for the customer's use.

The pharmacist handed Miller his purchase. "Will there be anything else?"

Miller ignored the man. He waved his datasleeve in payment, tucked the small package into a pocket, and walked out into the balmy Southern California sunset. Even in December, it was shirt-sleeve weather.

Despite the day's warmth, he shivered in anticipation of his SNAP experience. His respiration and heartbeat would slow to a nearly undetectable level. Blood at the surface of his body would plunge deep into its core to protect the vital organs. He would hover at the balance point between nirvana and death. In return for near-surrender to Thanatos, his reward would be hyper-cognition, an hours-long thunderclap of understanding.

Miller hurried eight blocks along Ocean Front Walk to his home, palmed the door open and ducked inside. An orange tabby cat curled around his legs mewling with impatient hunger. He hefted the cat and for a few seconds, the two nuzzled. Then the cat squirmed out of Miller's arms and yowled. It was past dinnertime and appetite prevailed over affection. While the mouser ate, Miller took his own meal, if six ounces of amino acids, fatty acids, and glucose could be called a meal. It appealed to none of his senses save hunger.

He walked through his modest bungalow to a plain bedroom, furnished only with a smartbed. He programmed it to maintain his skin temperature and ensure a comfortable recovery. He neglected this step once, and upon awakening, every centimeter of his skin burned with the devil's own pins and needles as warm blood returned to cold flesh.

Naked, trusting the smartbed to protect his skin, Miller lay down and activated the inhaler. He registered a brief tickle as billions of nanoparticles penetrated his nasal membranes. He could almost feel his brain flood with neurotransmitters. These chemical emissaries relayed messages to his body, barking orders to a fleet of corporeal agents. They slowed the nettlesome business of life support, system by biological system, putting vitality in nearly exclusive service to the mind. Miller was to be accorded a multi-hour experience of *satori*—Zen clarity without the fuss of *zazen* meditation.

At first he experienced the normal effects of SNAP. Seven seconds after inhaling, he felt his sinuses erupt and knew there would be a brilliant crimson trail where bloody mucus blanketed his face. The red stain was the source of the pejorative nickname: Rudolph. Then SNAP stilled his warming responses and he shivered. Even the hair on his body lay flat as the drug destroyed every source of thermal insulation.

But ah…the high! He was one with the cosmos—transcendent,

omniscient. He danced among the stars, sang the music of the spheres and soared along simultaneous paths of quantum particles.

The coppery taste was Miller's first warning that something was wrong. While he lay paralyzed in ecstatic thrall, blood began to puddle in his mouth. It rushed away from his core towards the skin's superficial capillaries, a torrent at escape velocity from the body's gravity well. It seeped from sightless eyes and deafened ears. It suppurated at a rate that would make hemorrhagic fever look like a bridal blush. Every centimeter of his skin oozed. It would be a race to see if he bled out or suffocated first. Five times before, NMech nanobots kept him alive. Today, he was swept across a biological Rubicon towards death's cold embrace.

Still, the body is stubbornly attuned to one lodestone, the irresistible pull of survival. This most powerful of instincts punched its mighty way through the chemical interference, demanding life for an unresponsive body.

All for naught.

Emery Miller often imagined that his final thoughts would be a flashing montage of his short life's events or that he would behold a mystical White Light proclaiming the Oneness of All. But Emery Miller's last thought before blood saturated his thousand-thread-count silk sheets and flooded his smartbed's sensors, before his heart stilled into silence, was to wonder, *Did I remember to feed the cat?*

❁   ❁   ❁

Three thousand miles away, in the sixth-floor management suite of a Boston office building, a chief executive sat at an ebony desk custom-scaled to fit her frame. A long bank of bare windows gave the space a clinical feel that matched the businesswoman's demeanor.

She'd scattered mementos on the opposite wall thinking this is what executives did, but the diplomas, photos, and a framed, jewel-studded gold pin were as out of place in the woman's barren office as a litter of puppies in an operating room.

A mat of dirty blond curls clung to her scalp like coiled worms. Her hands trembled, her legs kicked, and her eyelids fluttered uncontrollably. The 33-year-old face betrayed emotion for the first time in over a quarter-century. She'd pushed her body and mind beyond the limits that evolution had designed and her endocrine system rebelled.

Confused steps replaced her once-certain movement. Only days ago, her muscles had obeyed with a speed and precision beyond normal human capabilities, but now, on the rebound, she was riddled with tics and twitches. As she lost control within, she sought greater control in the world outside her.

Eva had a plan. The task was a difficult one, to create a master switch that would control every NMech product, every NMech customer. She was a scientist, so she would experiment. She would learn. She'd picked her test subjects carefully, as any good scientist would. Emery Miller was first on her list.

Miller had no family, no friends, no one to miss or to mourn him, none to question his death—a perfect test case. She peered into a heads-up display and then grunted in approval at her short list. Like Miller, the other three on the list lacked family or close friends. The soldier's entire world was his army. The scientist's was her career, and the tea expert's, his employer.

An electronic back door gave her control—not to the actual medical, recreational, military, and environmental nanoagents; any tinkering there would immediately be flagged to the systems that monitor product safety. No, Eva would control the *accounting* for these applications. It was simple: a bookkeeping entry thwarted all of the safeguards built into the company's products. It simply cancelled his life-support subscription for nonpayment. One stroke of a pseudo-accountant's pen had transformed Emery Miller from preferred customer to deadbeat, and then from deadbeat to…dead.

All NMech's products were rigorously tested to ensure the safety and satisfaction of its subscribers. But *bookkeeping* entries? Insignificant. They were of no more interest to the ardent sentinels of product safety than an ant would be to Cerberus, the three-headed beast that guards the gates to the underworld.

Eva Rozen's face twitched again, this time into a smile. Control was in her grasp. Cerberus was her pet, and programmed to do her bidding.

13

AN UNEASY ALLIANCE

BOSTON, MASSACHUSETTS

MAY 19, 2038, 7:00PM

When Jim Ecco told Marta about Eva's unexpected visit to the shelter earlier that day, he recounted her dramatic entrance, the receptionist's flight, and Eva's willingness to wait for Jim to finish his rounds with the dogs. Marta was skeptical, and at first refused even to listen to Eva's proposal. When Jim outlined Eva's plan to fund public health projects that Marta would administer, Marta immediately grasped the relationship between commercial nanomeds and paying for the costs of developing public health applications. But she was unmoved.

"I don't even want to hear the details," she said. "I don't trust her."

"Marta, if we can hold Eva to her word, you can save remedies that might be lost. You've said yourself that the cost to find the plants that have medicinal value, to isolate the active compounds, and then to synthesize the drugs means the pharmaceutical corporations are not interested in developing. Many of the remedies you've catalogued will be extinct before the drug companies think to fabricate them."

"I can't argue with that. But I haven't spoken with Eva in years. Why should I suddenly trust her now?"

"This is not the same Eva Rozen. And if she and you can manufacture with nanoassembly, you'll save some of the cures that will be lost if the rainforests die out. Maybe they'll recover, maybe not. But we'll have the medicines."

"And make Eva rich," said Marta brusquely.

"That's the deal. She gets what she wants, and you get what you want."

"What I want is for her to stay away from me. From us."

"Marta—"

Marta held up a hand to stop Jim. "Sorry. My answer is no."

"No? How can you say no?"

"I've been collecting plants since my first summer in Puerto Rico. There are still millions of hectares of rainforest to explore. I want to find and catalog what I can before the forests are destroyed."

"You sound more like a librarian than a scientist. At least people use the information that a librarian files away so neatly."

"That's not fair," Marta protested, but she knew that it was beyond fair: it was accurate. Worse, it was not what she wanted. She remembered Abuela's words, *"What becomes of adults? Do they follow their hearts or are they filled with discontent? Why not do what's in your heart?"*

Marta paced, considering the opportunity. Then she spoke. "Eva always says, 'I have a plan.' I can't see doing this unless we have a plan to make sure that Eva keeps her word."

Jim said, "You could insist on being equal partners."

"That's an idea," Marta allowed, "but I don't know if I can do it myself. I do like her idea of using commercial applications to fund public health, but I don't think I can do this unless you have a role at NMech, too. And not just some sinecure. I want all three of us to be involved. I can't square off with her by myself when we disagree. I don't do well with confrontation."

"But what about my work at Haven Memorial?" Jim protested.

"Querido, I know it's important to you, more than important. And so is my research into rainforest-based medicines. But you're right. This is a chance to do something that could change the world."

"I don't want to leave my training work behind. Leave me out of this," said Jim.

"And I don't want to abandon my exploration." Suddenly, both were breathing hard. Marta pressed on. "You've built a terrific staff. That's the mark of a good leader—your team could carry on without you. But I can't face Eva alone. If you become part of NMech, then we're two to one. Or she can have forty-nine percent of the stock and you and I split fifty-one percent so we have a voting majority."

"Why would she give me that kind of a role at her company?" he asked.

"Jim, think about it. You're great at organizing and leading people. I don't have much in the way of people skills—"

"Yes, you do," Jim interrupted.

"That's sweet of you to say. But I'm not a leader. I like to be gracious, but you have a knack for getting people to want to do the hard tasks. I'll consider working with Eva but I won't do it without you."

Jim sat and subvocalized and invoked a heads-up display.

He saw his son tinker with an old dataslate. Dragonfly monitors—insect-sized sensors that combined specks of processing power—produced visual images and sound readings in Dana's room. "What about Dana? Look at him. Eight years old and he's taking apart and reassembling his slate like it was a construction toy. Haven Memorial lets me set my hours so I can be with Dana. I want to stay connected with him. He's special, and I'm not just being a proud papa. He has a gift. What about that?"

Now Marta was silent. She paced their small apartment. The walls were drab, without the benefit of brightwalls—paint embedded with light emitting nanoparticles that allowed the walls, ceiling, and floors to provide variable lighting and heat at a command from a datapillar. There was no pillar, for that matter. Even something inefficient and underpowered was beyond the means of a family whose income was based on two soul-satisfying but low-paying jobs that kept a roof over their heads but little more.

She paused at a window. At least it had nanoglass, thanks to building codes rather than to any generosity on the part of their landlord. She touched the pane and it darkened. Looking out at an alley and a convenience store was dispiriting.

"There are afterschool programs at NMech," Marta said.

"How do you know that?" asked Jim.

"I follow things," she allowed.

Jim took a moment to digest that information. "So, let me get this straight. Eva knows about your exploration of rainforests in places like Brazil and Borneo. You know about kids' programs at NMech. You two haven't spoken in years, and yet you each know all about each other's careers?" Marta nodded and Jim shook his head. "You're like the boy and the girl in an old movie—fighting like cats and dogs throughout the film and lovey-dovey at the end."

"Don't count on lovey-dovey. I know about her because it's hard not to notice what she's doing. It seems like every scientific journal has a paper she wrote. Then there's the financial press. She spent four years at Harvard telling me that she was going to be the richest woman in the world and she appears to be well on the way. As far as her knowing about me, well, she spies on people."

"Is it any good?" asked Jim.

"Is what any good?"

"NMech's afterschool program."

"It's supposed to be. Apparently, Eva put some decent money into it," Marta replied. "You can look into it."

"What do I know about kids' programs?" asked Jim.

Marta smiled. "Guess you're going to find out."

"You mean you'll do it? Work with Eva?" he asked.

"Yes, but only if you join me, Querido."

Jim and Marta paced for a few minutes, mulling over the opportunity. They were excited. But the apartment was small, and they had to pick their paths carefully.

Finally, Jim said, "Will Eva want me as a partner?"

Marta looked at him for a moment, shook her head and smiled ruefully. "Men are so blind," was all she said.

After dinner, Marta took a deep breath and held up her hand to receive a file. "Okay, let me see what Eva has in mind." Jim reciprocated the gesture and subvocalized a command. Marta's datasleeve pinged receipt of Eva's proposal.

Marta invoked a heads-up holographic display and began to read through Eva's plan in stony silence. When she finished, she looked up at Jim, careful to keep her expression neutral.

She took another deep breath, muttered, "I know I'm going to regret this," and touched her commpatch and subvocalized a link to Eva. At first, her voice was controlled. Soon it became more animated and rose from an inaudible subvocalization to a clenched-teeth whisper, the commpatch equivalent of shouting.

She touched her commpatch and suddenly Eva's voice was projected into the room so that Jim could hear both sides of the conversation.

Marta was angry. Her voice dripped with scorn. "I want to attack morbidity, and you want to cure lactose intolerance? So that people can eat ice cream? *This* is public health? You want to cure *farting*?"

"Don't preach to me." Eva's disembodied voice shot out of the holograph that displayed her avatar. She sounded calm, as if she'd anticipated Marta's response. She said, "Just listen. I know that lactose intolerance isn't schistosomiasis, but who's going to pay for that kind of medicine? Who cares about any of that besides you?"

Eva continued evenly before Marta could interrupt. "You're right. Lactose intolerance? So what? *But people love milk* even though it's indigestible for most adults. Ice cream consumption alone is nine billion gallons a year. You're looking at a third of a trillion-dollar-market."

Marta broke in. She bit off each word and joined them into a staccato indictment. "I don't care about the money. Why did I link with you in the first place? I wanted nothing to do with you once we finished the Harvard project and now I remember why that was. You remind me of Humpty Dumpty—you make words mean what you want them to mean. You say, 'public health' and you mean 'get-rich scheme.'"

Eva's voice turned flat, matching Marta's passion with an affectless recitation. "Just listen. Have you ever known me to act without thinking? Ever? Just keep your mind open for two damn minutes."

"Of course, it's not a medical issue," Eva spoke with unchar-acteristic vigor. "That's why it's our starting point. *It's not going to be regulated*. You've got to see that. We can get it to market fast and make money—a lot of money. Then we can afford to do public health. I explained all this to Jim."

Marta shot daggers at her husband. "What else did you two talk about?"

"Give it a rest, Marta. That's ancient history."

Marta drew in a deep breath and exhaled slowly through pursed lips. "Okay, sorry. Go on. Say your piece."

Eva pushed on. "First of all, just to be accurate, there is no cure for lactose intolerance. Nobody ever thought it was important enough. Sure, you can spray an enzyme onto your food. Add lactase to the production. But those remedies are afterthoughts. Billions in ice cream sales alone and nobody thought to cash in? Wake up, Plant Lady. NMech rolls out a nice little consumer nanoapp that builds the enzyme. An effective remedy. And it's over-the-counter so that we don't get stuck in trials for years. Dammit, everybody likes ice cream and nobody wants gas. So we fix it. We go to market. We make money. Then you take part of the profit and go cure something. Anything. Whatever you like. That'll be your baby. But public health is a big investment and I'm not going hat in hand to some waste-of-skin bureaucrat for funding."

There was a long silence. Eva started to speak but Marta held up her hand to stop the woman's comments. The gesture was invisible, but Eva stopped in mid-sentence as if she were standing next to her.

Marta considered Eva's explanation. At last, she spoke again. "Eva, where did you get the idea for this?"

"From you."

"From me? I never spoke about lactose intolerance," Marta said.

"Didn't say you did."

"Then why do say you got it from me?"

Eva was silent. Marta waited. At last, she understood Eva's cryptic comments and flushed. "Oh my god. Why didn't you say something to me then?" When Eva didn't answer, Marta blurted out, "Was it bad?"

"Uh-huh," Eva said, stretching out the two syllables into simultaneous rebuke and exoneration. "Between you and Dana, when he needed to be changed, well, let's just be ladies about it and say that Mother Nature has an odd sense of humor."

"But that lab was tiny. It must have been..." Marta trailed off, embarrassed.

"Marta? You had enough on your mind back then. You had Dana. You weren't getting much sleep. And you were pissed about your dataslate and barely civil to me. And besides, I'm not big on girltalk. You think *I* can find a diplomatic way to bring it up? Uh, no. Not my style. You liked your afternoon scoop from Toscanini's. Plain vanilla ice cream, as I recall. But an hour later? Ewww. Gas. Every day. I never got used to it."

"Oh, my god," Marta repeated.

"But I can tell you one thing," Eva said.

"What's that?"

"If you still like your afternoon scoop? Then you're going to be our first customer."

❁   ❁   ❁

Six months later, NMech released its first consumer product, Easy-Milk, every bit the success Eva predicted. They followed up with FreeSkin, a nanoagent complexion cream that targeted acne in teens and young adults. It was an even greater success. Nine months after the release of EasyMilk, before NMech went public and started Eva on the path to becoming the world's wealthiest women, Eva delivered on her promise to fund medicine in the public interest. The uneasy alliance had worked.

Eva, Marta, and Jim met in the NMech boardroom. In keeping

with Eva's decorating sensibilities, the room was stark to the point of barrenness. A large oval cherry wood table dominated the room. The floor-to-ceiling drapes were set to a blood-red velvet, one that gave the room the look of an abattoir.

Marta subvocalized a command to the room's pillar and the nanofiber drapes reorganized into calm green linen. Eva turned at the change and stared at Marta. Marta smiled.

"I just set the drapes. I like red," said Eva.

"Well, good morning to you, too. Let's have something with some life in it. I feel like I'm walking into a whorehouse."

The two scientists regarded each other. Tension and the occasional contretemps seemed to be a permanent part of their friendship.

Jim walked in, too late to have heard the interchange. "Ladies, I feel like celebrating. Okay if I change the drapes to something festive?" Eva rolled her eyes and Marta managed a chuckle as Jim selected a display that depicted every recognized breed of dog. Marta took in Jim's decorating sensibilities and lowered her eyes.

"Isn't that better?" he asked.

"Oh, yeah, Jim, very classy," Eva said. Then she subvocalized.

The image shifted again and the room darkened as the drapes morphed to a black velvet. In place of Jim's bright array of proud canines, Eva had substituted the image of a popular though gaudy painting. Seven dogs sat around a green felt table littered with poker chips and cards. Each had a highball glass half-full of melting ice and whiskey. A collie on the left side of the table held her cards close to her chest. Opposite her, a terrier with a worried expression stared at his hand. The black Dane smoking a pipe looked smug and the bulldog held an ace in his left hind paw out of sight, below the table. A small brown dog with its back to the viewer appeared to be winning.

Marta and Jim looked up at each other and Marta subvocalized. The drapes returned to Eva's crimson.

Point and match to Eva Rozen.

❀   ❀   ❀

Eva's public health proposal surprised them. "I've got a great plan for public health."

"Are we going to talk about it?" asked Marta.

"What's to talk? I have a plan."

"I'd like to have a say in what disease we choose."

"Who said anything about disease?"

Marta's temper flared. "Eva, you gave your word."

"Yup."

"And now you're saying we're not going to work on a disease? No medicine?"

"What's so important about fabricating a medicine?" Eva, the Provocateur.

"What's so important?" Marta exploded. "Our whole point in joining you was to give help to the people who couldn't afford medicine." Marta, The Crusader.

"Wrong."

"You want to explain what you mean before I walk out the door?"

"Sure. You said public health. You, dear tolerant lady who always speaks so sweetly and never jumps to conclusions, are going to get public health. But nobody said we had to do pharmaceutical fabrication. It might not be too hard to assemble, but FDA trials and regulations would tie up our resources for years."

"So, what? You're going to give us an even better cure for gas?"

"Very funny. How about clean water?"

"What?" Marta and Jim said simultaneously.

"Yep. Good ol' H_2O."

Marta and Jim froze. Eva displayed a rare smile. She subvocalized and smiled again, and, in an uncharacteristic display, she laughed. Eva, the Surprising.

Marta did a double-take. Eva chuckling? Public health was momentarily relegated to an afterthought. "What's so funny?" Marta asked, incredulous now, rather than incensed.

"The looks on your faces. Priceless. I recorded it so I can watch it again whenever I need a bit of comic relief. Now let's think big, shall we?"

"Infectious disease isn't big? Fifty-five million deaths a year?"

"More like seventy million—a lot of it is uncounted," Eva said. "But one-quarter of the world doesn't have clean water. That's over two *billion* people who are thirsty." Eva stood and paced, hands clasped behind her back. Her shoulders hunched forward and she looked even smaller than her four-foot, four-inch frame. Her pale skin was a stark contrast to her black nanosilk cargo pants. She paused, and then turned suddenly. Eva chuckling, and now dramatic gestures? Jim and Marta stood spellbound, mouths agape at her sudden animation.

"And you think it's bad now? We're just beginning a cycle of even more severe water shortages, thanks to droughts from climate change, from over-pumping of underground aquifers and from relaxing clean water standards. Then there are all the construction practices that pollute drinking water." Eva, the Jeremiah.

She held Marta's gaze. "Compare clean water with inventing a new drug. Want to spin your wheels with FDA trials? And nobody's mentioned patent issues. What if our fabbed meds are indistinguishable from another company's test tube version? Suppose we assemble a drug someone else claims is theirs? You want to spend a decade with litigators over who owns the rights to what?"

"Come on, let's focus on what we can deploy immediately. There's a real need, Marta. Not to mention the commercial applications. I mean, if we happen to make a profit later on, would that be okay with you two Samaritans? Not to mention the publicity we'll generate."

"Eva, you are the most exasperating person I know." Jim, the Peacemaker. His voice was couched in tones of admiration.

"Just exasperating, Jim? Come on, say it: I'm the smartest, most wonderful person alive, and you worship the ground upon which I tread."

"Smartest, no question. Worship? Got to give that title to my wife and my son. But you are my oldest and dearest friend and I regard the ground upon which you walk on as something close to a national treasure. How's that?"

Eva said nothing. She felt warmth suffuse her—she was touched by Jim's affirmation of their friendship. She flushed.

So did Marta, the Insecure.

There was an uneasy moment as the atmosphere in the boardroom changed, like the stillness before a thunderstorm.

Marta drew a breath. "Do you have something specific in mind?" she asked, refocusing on the project.

"You ever know me to be unprepared?" Eva snapped and then continued without waiting for a response. "There's a water desalinization project in Venezuela that's perfect for us," She subvocalized and the room's pillar projected a globe onto the conference table. An image of the Americas and the Caribbean faced them.

"What are we looking at?" asked Jim.

Eva zoomed in. "This is the Paraguaná Peninsula of northern Venezuela, on the Caribbean coast. There's a desalinization plant there that keeps people in Central America and the Caribbean from dying of thirst. The plant can't keep up with the demand anymore."

"How old is the plant?" asked Marta.

"It went online twenty years ago."

"Why can't it produce enough fresh water for the region?"

"First of all, it wasn't designed to be the primary source of water. The principal source of fresh water in the Caribbean is rainwater, which is scarce during the best of times. After twelve years of the NAMSEA drought, reservoirs and emergency supplies are exhausted." Eva referred to the twelve-year water shortage that desiccated much of North America, the Mediterranean and Southeast Asia—hence the name, NAMSEA.

Marta asked, "How bad is NAMSEA? Compared, say, to the Sahel drought?"

"Which Sahel drought? 1910?" Eva asked, "The 1940s? 1960s? Or do you mean the 1970s? 1980s? 2000s?"

Marta frowned. "Uh, the bad one, I guess."

"The 1970s drought is the one most people think of. About 100,000 people died and millions were left homeless. That one?"

Marta nodded, still staring into holographic display.

"That drought was severe—for its time. The PDSI rating was bad, but not as bad as we're going to see," said Eva.

"PDSI?" Marta asked.

"Palmer Drought Severity Index. A zero means no drought, average rainfall and water table levels. Negative scores go higher as droughts increase. Sahel measured about a -4 on the PDSI scale. Right now, NAMSEA is running about -6, depending on where you measure it. Two years ago, you had water riots that killed over 20,000 people in cities across the Caribbean. We're starting to see dust storms like the 1930s. The dust smothers vegetation and destroys machinery. It shears off the arable topsoil. Not so good for agriculture." Eva paused and then continued, "Put it this way, I wouldn't put the Caribbean on my next vacation itinerary."

"Is it getting worse?" Jim asked, and then shook his head. "Dumb question. How bad will it be and can we really make a difference?"

"Actually, that's a good question. I've been following the work of the lead researcher at the National Oceanic and Atmosphere Administration. NOAA is staying mum, but privately, its people predict that the drought will hit -8 by the end of the century. That'll make the Dust Bowl look like a sauna."

Jim and Marta absorbed this information in silence.

Eva said, "It gets worse. The Puerto Rican government dredged harbors to make bigger port facilities and managed to damage natural aquifers in the process. Same thing in every island that fancied a nice new port. Mother Nature spends a few million years to carve water-bearing bedrock caverns. Commercial agriculture and mining ruin the karsts in just a few decades."

"Karsts?" asked Jim.

"Limestone caverns. Mama Nature builds 'em. We drain 'em."

"What happened?" asked Jim.

"Acid rain. Over-pumping. Who knows? Who cares? It's an opportunity for NMech. And the chance for you to do something in the public health arena."

"So how does this place—Paraguaná? How does it figure in?" asked Jim.

"Well, before the drought, the desal plant was the margin of safety in the Caribbean. Now with the drought, and with other water sources ruined, Paraguaná can't make up the difference."

"What's wrong with the plant?" asked Jim.

Eva said, "The key problem is that they use reverse osmosis."

"We've been using RO for something over a half century. Longer even. Why is that an issue?" asked Marta.

Eva explained. "Old technology that made sense when energy was cheap. But you need a hell of a lot of energy to force water through a filtering membrane. Now it's too expensive to filter water that way but they're stuck with the plant. But the biggest problem is that RO works too well. It's self-defeating. Water slips through the filter but salt and pollutants build up and clog the filters. The more water you try to process, the more the sludge builds up till it fails. Ironic, huh?"

"What's the solution?" asked Jim.

"Nanomembranes," said Eva, with an open hands gesture, as if the answer were so obvious a child could give it. "Nanopores made from carbon or boron nanotubes are about 50,000 times smaller than a human hair, but will process more water, by several orders of magnitude. The material is so slippery that the buildup sloughs right off. It's like the difference between a cocktail straw and a fire hose. And since water races through the pores with almost no friction, the plant will be able to lower its energy consumption."

"Can NMech do it?" asked Marta. "This would be fantastic."

"Sure can. In fact, I have a plan for it."

"Of course you do!" Even Marta laughed, caught up in her colleague's enthusiasm.

Eva continued, "Look, we can build the filters in an assembler. That's easy. We'll have Paraguaná at capacity in four months. The initial yield may be ten times greater than what it is now. And the best part? We have a demonstration plant. We can turn around and sell the technology to industry." She looked pointedly at Marta. "Are we all happy girls now?" then winked at Jim.

"What about logistics?" Marta asked, serious again. "Once production rises, can the existing pipelines handle the increased output?"

"Good point. We'll need to make some upgrades but I think with the proper management, there'll be enough transport capacity by the time the plant is ready."

"Is that realistic?" asked Jim. "There must be thousands of miles of pipelines to carry water. If we up the production by a factor of ten, are we going to be able to get the water where it's needed? It's like expecting, oh, I don't know…like expecting a bike path to handle a highway's worth of traffic."

"Actually, it is realistic. When the plant was built, the promoters overstated what RO could produce. Overstated it big time. So there's been excess capacity since day one. We'll need to do some building but nothing extreme." Eva paused and looked to make sure she had their attention. The side of her mouth curled up in a half grin. "Besides, NMech just purchased the two local suppliers of pipeline and fittings so there'll be some additional revenue."

"Of course we did," Jim laughed. "Great plan, Eva."

The compliment cracked Eva's impassive expression and a smile stole across her face and hovered for an instant like a hummingbird at a flower. A second later, her features resumed their characteristic impassivity.

"Well, I'm in, I guess," Marta said. "But I want to get back together in, say, six months and look at what we accomplished. Lessons learned and all that."

"Okay," said Eva. "Anything else?"

"One thing," said Marta. "I want you to meet an old friend of mine. She's the developer of morphing nanocouture. I think there might be a place for her at NMech."

"We have our own textiles division."

"Exactly. But that's military textiles. They're tough and self-repairing, and the uniforms don't change much. So that division is becoming less profitable. Think of nanocouture as cash flow. Styles change, and that means constant new business." Marta smiled, her excitement evident at the opportunity to present a business case. "It also means that we'll have a peaceful use for some of our military technology, and that's important to me."

"Well, well," said Eva. "Listen to who's talking like a business-woman. I like that. So you've got, what, a model for us? A fashion queen?"

"No, a scientist who's interested in fashion. She's an old friend of the family and I was waiting for just the right moment to suggest she join us."

"You've known her long?" Eva sounded half-interested.

"She's a close family friend. In fact, Dana calls her Aunt Colleen. But I'm suggesting you talk with her because of her science, not our friendship. Morphing couture can be huge, but she doesn't have the capital to develop it and to compete with the established fashion houses."

"What the heck is morphing couture?" asked Jim.

"It's a way to use datasleeve and software rather than needle and thread. A command to a datasleeve and Colleen's pieces recon-figure into various styles. Right now, you can only change the color and texture of a garment. With morphing couture, a single garment changes into a designer's newest styles. It'll be like nano-custom-ized prêt-à-porter," Marta said, her voice rising in excitement. "The styles can even be programmed to expire when each new line comes out. That means customers make multiple purchases."

Marta said to Eva, "I told Colleen to meet me here so I could to introduce you two. She's got a PhD in nanotextile materials engineering from Harvard. That's where we met."

Eva shrugged and returned to her model of the Paraguaná RO plant. Marta subvocalized and a moment later the boardroom door opened.

A young woman entered the boardroom and smiled. A lush cascade of auburn hair in a loose braid served to accent her slender neck. She'd brushed delicate metallic streaks into her hair that projected tiny electrical emissions like a subtle halo. They glowed and flickered to draw attention to an elegant face. Her features were precisely symmetrical—full lips, captivating green eyes, and an aquiline nose. The distance from eyes to lips formed a pleasing proportion.

Eva looked up and gasped involuntarily. She was looking at a woman who could have doubled for Gergana. The wide smile, the innocent eyes, the full hips, the perfect facial features. Eva blanched.

"Eva," Marta began, "I'd like you to meet Dr. Colleen Lowell."

Colleen stepped forward to accept Marta's introduction. "I'm very honored to meet you, Dr. Rozen. You're one of my heroes, an inspiration to women scientists, and—"

Colleen stopped and stared at Eva. "Dr. Rozen, are you all right?"

Jim and Marta turned to their colleague. She grimaced, as if in pain, and sat down heavily. She rested her elbows on the cherry wood conference table and held her head in her hands. She looked up again at Colleen.

"No," muttered Eva, struggling to regain her poise.

"What's wrong?" said Jim.

Eva looked at the bewildered trio on the other side of the table. She repeated, now in a firm voice, "No." Then Eva rose and stalked towards the boardroom's door.

"Eva?" Marta asked, concerned and confused.

Eva said nothing. She paused for a moment, and then turned back and scrutinized Lowell.

"No," Eva said, for a third time.

"What the—?" Marta said. She turned to Colleen Lowell and spilled out an apology. "I'm sorry, Colleen, but our partner, uh, Dr. Rozen, she can be eccentric every now and then."

"What's going on, Marta?' Lowell asked, an edge to her voice. "You said that NMech would take on my work."

Before Marta could answer, Eva turned to the young woman. "Settle down, Ms. Lowell," said Eva, in her customary flat voice. She had regained her composure.

"That's *Dr.* Lowell, if you don't mind," Colleen huffed. "Materials engineering at Harvard. I believe that's your alma mater."

Eva stared at her, unblinking, and then said, "Marta, your idea is a good one. I do not want her up here in the executive suite. I do not want to come in contact with her." Then, turning back to Lowell, Eva said, "Nothing personal."

"Not personal? You tell me you don't want to see me and that's not personal? Well, it's personal to me. I don't care if you're the smartest woman in the world. You dress like you're going to a cattle auction—as the livestock. I deal with people in the fashion world who would eat you alive, you ugly runt! Marta, thanks for the invitation. Good luck with Raggedy Ann."

The words, *ugly runt,* triggered a flood of memories for Eva. Others had hurled the same words. Mama did. Papa did, especially when he believed she did not hear. Schoolmates did. And Bare Chest did.

Each taunt cried out for recompense and Eva kept a ledger. Recording each offense was an automatic mental process, if not a conscious one. Each offense was tabulated with meticulous precision.

Eva remembered some of the offenses. There was the child— could she have been six? Seven?—who uttered those words. That child soon watched a shiny thing, half-hidden in Eva's left hand, turn into a pair of scissors. Then she saw a hank of her pretty red hair fall to the playground tarmac…the boy, who returned from soccer practice shower to find every piece of his clothing glued together…the schoolmate who found her hair conditioner replaced by a depilatory.

The foulest and most memorable entry in her ledger was that of Bare Chest—Akexsander Yorkov—accomplice to Gergana's murder. Eva remembered him clearly, and the price he'd paid.

Colleen's angry retort may not have triggered a conscious memory of each violation. But it evoked an emotional response from a lifetime catalog of insults. This triggered a series of biological events. Eva's endocrine system prepared her for fight or flight, flooding her with adrenaline and noradrenaline. Her heart raced and digestion slowed. Glucose pumped into her bloodstream and her pupils dilated.

Years of free-flowing rage found a target, not two feet away.

Eva smiled.

She touched Colleen on the forearm and spoke quietly, even soothingly. "Be careful, dear. You meet the same people coming down as you did when you were going up." Eva held her hand on Colleen's arm for a moment longer than necessary to make her point, and a series of software commands flowed from Eva's sleeve into Colleen's. She did not need to touch Colleen for the rogue software to jack into Colleen's sleeve but the unexpected contact distracted the fashionista from Eva's real purpose

Lowell jerked her arm away and stormed out of the boardroom.

"Well, that worked out nicely," said Marta. "Thanks, Eva. Real professional. That was a very dear friend you just insulted. And one heck of an income stream you just threw out."

"Disregard it. We have enough on our plate right now," Eva snapped. "And nobody calls me that name to my face."

Marta wanted to say, *Which name? Ugly? Or runt?* but she held her tongue.

Eva left the boardroom fighting to control her emotions. Something tugged hard at her memory, but it was frustratingly out of reach. Eva's breathing became shallow, rapid breaths that drew in little air. She felt her pulse throb in her neck.

Eva was processing images, memories from her preverbal infancy. Lowell's sculpted features matched Eva's stored images of her sister and her onetime caregiver, Gergana, both alive and dead in Eva's unconscious mind, where past, present, and future were indistinguishable.

Ego structures strained under the reanimated weight of memory and loss. Long-repressed images pushed insistently against the barriers that separated a violent past from a controlled present, like protestors overrunning a police barricade.

The din from the Table of Clamorous Voices had been dormant. Now it was an unquiet phenomenon.

14

 HOME SCHOOLING

FROM THE MEMORIES

OF DANA ECCO

I was almost nine when my parents bought a home in Pill Hill, an upscale Boston suburb, named for its proximity to a cluster of hospitals. My father cycled to work, weather permitting, or rode with my mother in a P-car, a semi-private driverless automobile. Our house backed up against a ribbon of parkland called the Emerald Necklace. That was my western frontier, my Sherwood Forest, my mystical kingdom. If I wanted to meet Robin Hood's Merry Men, cowboys, or elves, I had merely to walk out the door and into my backyard imagination.

My mother showed me plants in the parkland that were edible or medicinal. Today, when I walk through the parkland along the

Muddy River, I can still spot trillium, wild asparagus, leeks, and Solomon's seal.

My childhood was as ordinary as anyone raised by one of the greatest scientists of her age and tutored by another. I liked to listen to music, to play games—things adolescents have done for centuries—although sports never caught my fancy. I tried to play soccer a few times, and lacrosse just once. Evidently, an easy-going personality doesn't match up well against big-boned bloodlust.

My playmates were more inclined to fantasy than to football. We found our castles, battlefields, and alien landscapes along the Muddy River. The fens and woodlands were forests and jungles inhabited by wild creatures. What courage, what daring we displayed.

I made friends with children of other scientists at NMech, explorers all. We hunted for treasure and found it, right there at NMech—ancient sensors, induction coils, spectrometers, rheostats, and voltmeters. I doubt that another school, private or corporate, boasted a scanning tunneling microscope, capable of nanoscale resolution, right alongside a considerable pile of building blocks, board games and baseballs.

Eva Rozen took an interest in my adventures. She listened to my tales of gallantry and adventure without once censoring me. I didn't realize that she thought made-up stories were a fool's task until I was much older.

Our family life was much like that of other families. I got along with my mother and father, debates over household chores notwithstanding. After my mother became a major stockholder at NMech and she hired the housekeeper I'd lobbied for, we found other duties over which to disagree.

I also lobbied for a dog, but my father surprised me and refused.

He said that we'd be away from home too much. When I reminded him that we could take a dog with us to NMech since we owned part of the company, he smiled and said, "I'll think about it"—the universal parent code for "no." I think that after Ringer, he decided not to face the certain prospect of loss with another dog. Instead, he poured the part of him that craved primal connections with canines into his devotion to the dogs at Haven, and to my mother and me.

Schoolwork was easy until my parents decided to take over my education. Homeschooling was harder, but more enjoyable. My father, my mother, and Eva were my main teachers. Life sciences, like biology and botany and medicine, how the body works, how nature works was my mother's area. I learned about the Taíno, about Abuela, but I wasn't to meet her for a few years.

My father tried to teach me social sciences, but he'd ignored those subjects in high school. So it was up to me to learn at my own pace. Instead, we tackled literature and the arts together. We read a lot of the old mystery and science fiction stories—Conan Doyle's Sherlock Holmes, Edgar Allen Poe's short stories, Raymond Chandler, Asimov, Bradbury, Frank Herbert, and Julian May, for example—and called these the "Classics" to satisfy one of my educational requirements. He taught me what he called the people sciences: how to read body language and predict what people would do, how to pick up clues from their habits and grooming that would tell me about their lives.

My parents could monitor me when I was home or at school through the ever-present commpatch—until Eva showed me how to jack the patch so that I could choose background sounds for them to hear. Then I had a great deal more freedom.

During the years before the Great Washout, Eva was as much a mentor as my mother. I studied chemistry, physical sciences, and computer science with Eva. I thought it was the most natural thing that the soon-to-be-richest woman in the world spent time nearly every week with the juvenile son of her business partner and scientific colleague. Our studies went a bit beyond the traditional sciences. Again, I thought it perfectly natural for my extracurricular activities to be preceded by the warning, "Don't tell anyone".

It was always "Eva." She had said, "You're like family for me, but don't give me any 'auntie' or 'sister' or 'Mama Eva' crap. I'm just Eva. Got it? Maybe someday that will change. Maybe I'll adopt you," she laughed, "and then one day all this will be yours." That was a running joke with us, especially when an experiment or a project failed.

So I called her Eva, as if she were an equal, not a teacher, and without realizing that laughter was a rare display and a precious commodity for her. I think she wanted a playmate, a child who could understand her and get excited by the things that made her excited. I can't imagine her playing with other little girls when she was young. Dolls or tea parties would not have been within her repertoire. I doubt other children were interested in the periodic table of the elements or the Standard Model of particle physics.

I once asked Eva why she liked to play with kids—meaning me. She told me that when we spent time together her mind was quiet. Even so, there'd be times when she'd be distant, mute, and seemed to move under a terrible weight. I imagined that a giant hand pushed down on her. At first, I thought that's why she was so short.

She was an exciting teacher and companion. She taught by telling the stories of scientific advancement, which was strange because she considered storytelling to be "worthless nonsense."

The first story she told me was about Richard Feynman, the physicist who started people thinking about what would come to be known as nanotechnology. In December of 1959, he made a famous speech at Caltech and offered a prize of $1000 to anyone who could reduce a line of text to 1/25,000 of its original size. The scale was an inside joke: that was the reduction needed to be able to fit something called an encyclopedia on the head of something called a pin. I think that the encyclopedia was some kind of book of knowledge, or maybe it was several books. A pin was a fastener with a sharp point. I didn't quite get the connection, but it was well-understood nearly a century ago when Dr. Feynman issued his challenge.

When a scientist tried to claim the prize, Feynman almost couldn't pay. He didn't expect anyone to succeed for years and was hard-pressed for the funds. "But that's science," Eva explained. "It moves a lot faster than people expect. Tell a well-educated idiot what science can do right now and he'll call it science fiction."

Not only was it hard for Feynman to pay, it was tough for the winner to claim his prize, because the text was so tiny compared to the relative size of the pin that the scientists had to search to find it. "If you ever want to hide something," Eva told me, "you can leave it right out in the open. In fact, it's harder to spot in the open. Just make it very, very small—nano-size."

Eva told me the story of Feynman's challenge many times and the lesson stuck with me. I liked the idea of hiding in plain sight. As it happened, this lesson mirrored another. My father and I had just read Edgar Allen Poe's story, "The Purloined Letter." Poe's detective, Dupin, finds a letter that several people searched for but missed. The letter was left in plain sight, and overlooked, instead of in a hidey-hole that would have been searched. My father, with his ability to read people and the tiny details surrounding them was like Dupin, or Sherlock Holmes, who solved mysteries with the tiniest clues that nobody else could see until he pointed them out.

The lessons about hiding in plain sight would prove fortunate. Eva was at her most exciting when she was ghosting—I think that the old term was hacking. She travelled through private cloud data like a hungry barracuda swims through a school of minnows and she was just as dangerous. I'd seen her track people she thought had insulted her and play havoc with their pillar or sleeve. I was a willing pupil and occasional accomplice—I enjoyed mischief as much as any kid does—but she could be mean. I didn't like being with her then. The ways she got even were amazing, but if you were the target, you wouldn't like it one bit.

Eva also taught me to keep a journal of what I learned. She kept one and told me that every good scientist keeps a journal. If you found her journal, you could read it—if you had good jacking skills. But you would have to know about advanced chemistry and nanotechnology just to understand it. And you would need a great deal of imagination to visualize one of her plans, and even more courage to contemplate it.

15

COUNTERPOINT

FROM THE MEMORIES

OF DANA ECCO

Late one summer afternoon, Eva surprised us when she displayed a genuine smile and announced, "I'm going to take you out to dinner."

My mother's hair was drawn back in a loose ponytail. I could see her face register mild surprise and then incomprehension. Her eyes widened and her eyebrows arched to a peak well above the midline of her broad Taíno forehead.

"Why?" my mother asked.

Eva did the unthinkable. She smiled again. "No agenda," she said quietly. "Hungry, maybe?"

"Oh, boy," my father grinned and rubbed his hands together like a child rolling strands of clay spaghetti. "*Merci*, Dr. Rozen, *mon amie,*" he said with a contrived French accent. "Where are we going?" He sounded like a cross between a poodle and a Chihuahua with phlegm in its throat.

"North Shore. Company car and driver's waiting. Come on, Marta, relax. When was the last time we had a friendly outing?"

"Not in a long time," my mother conceded. "If ever," she added under her breath.

If I heard her, then Eva did, too.

Eva snorted. "Oh, you kidder," she deadpanned though her smile remained. "How about you let down your hair tonight?"

My mother looked startled and was about to reply when my father stepped between them, turned to Eva, and said, "Sounds great. You buying?" Eva nodded, with a brief roll of her eyes. "Then let's go," my father said. He turned to my mother with a smile, offered his arm, and said, "*Mademoiselle?*" This act of exaggerated gallantry defused the tension. Or maybe it was the ludicrous attempt at dialect. My mother took his arm and smiled at Eva. Her arched eyebrows settled first into the facial equivalent of parade rest, then, at ease. Eva's face returned to expressionlessness. The strain that normally bound the two abated. She took Eva's hand.

They were warily rebuilding their friendship. There were clumsy moments, like a musician stumbling over a difficult passage in a work that had lain unpracticed and the muscle memory lost. They were still friends when they moved from Los Angeles to Cambridge, Massachusetts, to study at Harvard University. Maternity and a crushing pre-med courseload demanded all of my mother's strength. Eva seemed to skate through her courses, a facility that likely nettled my mother. Perhaps it contributed to their falling out.

The driver opened the door for us and we piled in. My father got in the car and bounced on the resilient car seat a few times after the driver closed the door. He rubbed the seat covering and murmured, "Ooh. Could this be this real leather?" Then to the driver, "Are we sitting on cows?"

"No, Mr. Ecco, nanofabrics."

"I wonder how many atoms had to die for us to have this luxurious ride."

"And I wonder how many times I have to hear that tired old joke," my mother said. She smiled and accepted the festive character of the day.

My father played with the various passenger controls. The air flow stuttered on and off, while the music alternated among disparate genres.

"Jim, will you please sit still? You're worse than a child." She looked at me and said, "Tell me you're not going to grow up like that."

He just smiled and continued to play. The many gizmos he now had at his disposal at NMech seemed to help him compensate for the rigors of New England life. He had been raised in Southern California shirtsleeves and never adjusted to the extended cold of northern winters. He complained every time he offered a friendly greeting to a passerby and was met by downturned eyes. A hidebound Puritan legacy had gripped Boston for four hundred years: "Keep your eyes down, mouth shut, and thoughts hidden."

"Where are we going?" my father asked, as he settled down.

"Fine dining," Eva said. "Nothing but the best."

"I'm not dressed for anything fancy," my mother complained mildly.

Eva bit off a retort. Instead, she replied, "I mean, fine dining as in good food."

That evening her eyes afforded me an intimacy she seldom shared. The vulnerability was ephemeral and genuine. I could see a panorama of torment and joy—her madding fight for survival and the orderly structure of science in which she took refuge. My father, despite his uncanny abilities of observation, never seemed to notice, nor did my compassionate mother respond to this damaged woman's concealed disquiet.

Decades later, at the funeral, one that was shunned by all except a few members of the media, her eulogy included a quote from an earlier century's actress, Audrey Hepburn. Eva, the officiant said, "was born with an enormous need for affection, and a terrible need to give it."

It was an odd comment, given the context, but accurate. My birth gave Eva an outlet to express herself in a way that would have otherwise been impossible for the driven woman. We forged a curious bond. She was both playmate and mentor. My father told me that when I was an infant, she and I relished endless rounds of peek-a-boo. "Hello, Baby!" she'd call out, a bit too loud, and startle, then delight me. Otherwise, Eva never spoke to me except as she would to another adult. When we made up songs together—that was not her strong suit, it was too much like made-up stories—we were never quite able to work 'graphene' or 'quantum particle' into the lyrics and rhyme scheme of a child's song. We didn't care. We had fun.

As I grew, we competed. Our favorite contests were insults and math games and by the time I was ten or eleven, I seemed to hold my own with both. It's hard to imagine a juvenile matching wits with one of the great minds of the time, but it's equally hard to imagine Eva choosing to forfeit any competition.

There was one off-key note in those wonderful years. I think that my mother sometimes felt eclipsed. Eva held a role something like a grandmother and a grandmother figure evoked the pain and the loss my mother felt when her own mother died. Her eyes might glisten just before she issued an edict to end whatever game

Eva and I had invented. "Dana, time for your bath"—or dinner, lunch, snack, homework, chores, or an errand for which my help was suddenly indispensible. Eva would give me a sly smile, as if to confirm the temporary nature of the interruption and then she would turn back to granite.

The car glided silently to a stop at a roadside stand in Revere, a seaside town just north of Boston. The eatery was famous for its fried clams and the aroma of fresh oil and the sea drew a hungry mob. They milled about the same service windows that greeted customers for close to a century. People pushed their way up to the front of the throng to order and then drifted back to wait for their food, and then returned when their dinners were ready. They were like geese in flight, a few birds flying at the point to carve a path in the atmosphere, and then moving back to rest and draft behind the skein before pulling forward again.

We took our food and walked across a pedestrian walkway to sit on the sea wall. The broad ribbon of concrete unrolled along the three-mile length of the beach. We relaxed, ate, and watched people strolling past. My mother seemed engrossed in the ocean, lit now by low-angled rays as the sun set behind us. The colors that dappled the ocean's surface changed with the trajectory of the setting sun, violet and blue streaks giving way to yellow and orange, and finally blood red, the last wavelengths of the sun's declension.

After the sun set, an onshore breeze chilled us. As soon as we were back in the NMech car I kicked off my sandals and stretched out along the bottom segment of the sofa-like seating area. My mother and Eva bracketed me, each sitting on opposite banquettes, uprights of the U-shaped passenger area. My father was next to my mother, engrossed in a holographic depiction that only he could see.

Eva selected some music, Bach's Goldberg Variations. The simple aria that begins and ends the composition stands out in my memory. Even today, Bach's melodies take me to a place of peace, and the counterpoint takes me to one of balance. My mother stroked my forehead with idle affection. I looked up at her face. It was framed by her sable hair and a slight smile caused her eyes to sparkle.

I felt another hand. Eva gently tickled my feet. Twin caresses bracketed me, like the music's counterpoint. I floated in the music and the satisfied exhaustion of a day well-lived and hard-played. Then both sets of hands froze. I looked up and saw that my mother's and Eva's eyes were locked, one on the other, each with a gaze that held equal measures of compassion and possession.

My father must have noticed. He collapsed his heads-up display, reached out, and took Eva's hand. He gently pulled her across the car to sit next to him and placed his other arm around my mother's shoulders. All three looked contented—even Eva. I nearly laughed at their chained embrace, eyes closed and heads tilting back, resting on the car's soft headrests. They were three dolls posed on a shelf for sleep by a child taking good care of her playthings.

The only sounds in the car were Bach and the slow, synchronized breathing of three friends, who were at peace. Anything seemed possible except failure.

16

 ZVI

NMech's first foray into desalinization was a success. In a few short months, output skyrocketed at the retrofitted Paraguaná desalinization plant. Now Eva, Marta, and Jim were back in the boardroom, and once again vying to choose a color for the drapes. This time, Eva and Marta's debate was relaxed, even diplomatic. This time Jim prevailed and the drapes showed a peaceful view: a bloom of jellyfish drifting in an endless ocean.

"Well, we're heroes," Eva began. "We got water to the masses and kept several nations from civil disorder, not that those countries would have even noticed a regime change."

"Are they safe now?" asked Jim.

"Long as the plant keeps operating. But shut the spigot and I guarantee you'll see some big time civil unrest down there."

Marta said, "I'm proud of this one. I think we did well. Eva—thank you."

"You're happy, Marta, I'm happy." Eva nodded slightly and with gravity, as an empress to a countess. "Now we can take this further. There are several commercial applications we can focus on."

"What do you have in mind? I admit I had my doubts, but you pulled this off. What's next?" Jim asked.

"Kidney dialysis for one. The Holy Grail of dialysis is an internal device instead of patients being hooked up to an inefficient dialysis machine for several hours per week. I think what we learned at Paraguaná can be applied to build an implantable dialysis device."

"Sounds interesting," Marta said. "How do we fund the public health part?"

"Fund it? We're not going to fund anything. If people want to live without spending time in dialysis, then they become customers. The manufacturing costs are low enough that most people will be able to afford the gizmos. There are surgeons' fees, but that's not our concern."

Marta spoke up. "I'd like to make that our concern. Paraguaná was supposed to be a public health project, but we're going to recoup our costs with the commercial applications we're licensing for desalinization. There's money left in the pool we created from EasyMilk profits. Let's take some of that cash and use it on dialysis for the hardest hit populations. I don't mean we have to pay the bill for everybody, but I'd like to donate enough so that we can help, say, the poorest ten percent of renal failure patients."

"I don't think so," said Eva. "You wanted public health, I gave you Paraguaná. The fact that we parlayed that into profit is irrelevant. We can give away some of the devices but I'm not paying any surgeon's fee. I'll put my own grandchildren through college, not some rich doctor's."

"Give me a break. You don't have any grandchildren. You organized the desal plant and you did a great job. But you also found a way to get massive publicity and public good will. You made the good deed profitable," Marta said.

"What's wrong with that?"

"Not a thing," admitted Marta. "And I'm sure you'll find a way to make this profitable too. All I'm saying is my charter is public health. I want to work with the poor."

"Why them?"

"Because they'll die if we don't."

"Why is that my problem?"

Marta glared. The pacific mood in the conference room turned stormy. "Well, look at it this way, Little Miss Charity. Say you keep an extra 10,000 people alive. They're tied to NMech through dialysis. Wouldn't most of them become NMech customers for all their medications? Then you can turn a profit on them."

"Good point." Eva missed or ignored the sarcasm. "Let's see if the numbers back you up." She invoked a heads-up display and peered into it. "It'll take about two to three years for a charity customer to generate enough revenue with other purchases to recover the cost of implantation. That's a bit long for break-even, but there's the increased life expectancy from the dialysis. That should cover it. Okay, Marta, bring on the masses."

"Just like that?" Marta asked. "What's the catch?"

"There's no catch. The numbers add up. If this is what it takes to keep you happy and continuing to find cures in the jungle, then that's what it takes."

"They're not jungles, they're rainforests." Even conceding to Marta, Eva managed to provoke her.

"By the way, Marta, your thinking is good but your math is off. Helping 10,000 people is on the low side. Think of the recipients as an investment. Couple years to hit break-even, and then each one is profitable. Think big."

Marta stiffened, but Jim broke in. "Wait! Aren't you two forgetting something?"

"What?" demanded Eva.

"Uh, don't we have to *develop* this little invention first? I mean, nephrologists have tried for decades. Shouldn't we set aside a couple days next week to invent a device that's eluded science for the last half century?"

"Why Jim, now you're starting to sound like your wife. Anyway, I think this is a bit closer to her expertise. You want to organize a research team?" Marta agreed after reconfirming that the project would include a public health component. The three reviewed the basics of what they would need to start and agreed to meet again to discuss strategy further.

The tension had evaporated in the boardroom and the three colleagues enjoyed a respite from quarrelling. EasyMilk and Free-Skin were stunning successes. The simplicity of the desalinization project, coupled with the scope of its potential benefit, had won even Marta's trust.

Almost as a lazy afterthought, Jim asked, "Well, Eva, once we conquer every known disease, what's next?"

Eva said, "I've got a bigger plan."

"What's that?" asked Marta. Flush with the success of Paraguaná, Eva could have suggested a time machine, immortality pills, even a cure for the common cold, and Marta and Jim would have taken up the cause. But her next brain child caught them by surprise.

"Ready? One word: remediation."

"What the heck do we know about environmental cleanup?" asked Marta.

"What difference does it make?" said Eva

Marta stared and raised her eyebrows. She lifted her hands, palms up, in a 'What do you mean?' gesture.

Eva said, "Relax. We develop a good plan and we go to work."

"But we don't know anything about cleaning up toxic waste. I don't even know what we don't know," said Marta.

Eva waved off Marta's objection. "It's just basic chemistry. I found the perfect project we can start with. It's huge, and when we pull it off, we'll be the leader in remediation. Ever heard of the Nuovo River? In Rockford, Virginia?"

"No."

"Well, pretty soon the whole world will know about it. This one project could establish us as the hands-down leader in nanotechnology and put us into the remediation game—big time. Think about it. It fits right in with public health."

"I agree with you, Eva," said Jim, "but I also agree with Marta. We don't know anything about remediation. I'm not sure anyone will take us seriously."

"That is the major challenge," Eva admitted. "But we can sell it. And if we can get our hats in the ring, we can do it."

"Well," Jim said, "your track record is pretty darned good, Eva. It's hard to argue with success. Right now, I bet you could sell lies to a politician." He and Marta chuckled.

Eva did not so much as smile. She had already dismissed the bucolic splendor of the Nuovo River, which many considered to be among the most beautiful in the world. She dismissed Rockford's plight, a town wedged between environmental woes and economic concerns. Rather, she was impatient to see a stinking, nine-mile toxic stretch of water in southern Virginia where the river hugs the Rockford Munitions Plant. Into this aquatic embrace, the plant spews out pollutants and turns the watercourse from pastoral to poisonous.

The gunk had built up for close to a century, increasing and abating with arrival and departure of war. Eva's attention was fixed on the Pentagon's budget not the river's despoiled beauty. When the military bowed to public pressure and announced that it would seek proposals for cleaning the site, Eva began to plan. Military business was big business. The textiles division sold armored uniforms to the military, but its profits were marginal. Rockford would be huge.

"So, what can you tell us about this project?" Marta asked.

"Not a lot. The military classified much of what was dumped into the Nuovo, but we're pretty sure that there's dioxin, mercury, lead, ammonia, and copper. Not to mention DBP—di-n-butyl phthalate—which causes fetal mutations."

"Lovely," said Marta, wrinkling her nose. "The stuff we know about includes heavy metals, carcinogens, and mutagens. Am I getting this?"

"Yup. There's lots of it."

"And then there are pollutants that the military refuses to identify?"

Eva nodded.

"And we don't know anything about remediation?"

Eva nodded again. "No more than an undergraduate science major would."

"Sounds perfect. Let me guess. You have a plan?" Marta asked.

Eva nodded. "The cleanup shouldn't be too hard. The technology is mature. Use ZVI and scoop."

"Huh?"

"ZVI. Zero valent iron. That's iron in its pure state."

"Iron?" A note of incredulity crept into Marta's voice.

"Not just iron. Pure iron—zero valent iron. Doesn't matter too much what's in the water. ZVI takes it out."

"Can you expand on that a little? Dioxin, I understand. It's gotta be among my favorite industrial wastes. Who wouldn't love a carcinogen? But I wanna hear about this, this…iron stuff."

Eva ignored Marta's sarcasm. "Okay. Start with iron's instability," Eva said.

"Give me a break. Iron isn't unstable. What are you talking about?"

"Look, I don't tell you how flowers grow. Don't lecture me about chemistry."

"Well, excuse me. Why don't you explain how iron is," Marta made quote marks with her fingers, "unstable."

"Actually, in its pure form, it is very unstable. Iron has two, or sometimes three, electrons in its outer shell. It wants eight to be complete. So it wants atoms with electrons to spare, or to donate the electrons in its incomplete outer shell. It yearns to combine with other atoms. That's how it binds with the contaminants. The pollutants become chemically locked to the iron. The iron-heavy sludge settles out of the river. Scoop it up, haul it away and, bingo. Your cesspool becomes a swimming pool."

"Electrons yearn?" asked Marta. "Do they write poetry, too?"

"In a sense they do yearn," said Eva. "They're driven, compelled, motivated—you pick the word that makes you happy. But atoms want their outer orbits to be complete. So they either shed a few electrons or grab a few."

"Interesting," said Jim. "So, how come nobody's done this before? After all, iron isn't exactly a rare metal."

"You're right. It's maybe the tenth most common element in the universe. The problem is one of logistics. How do you keep the

ZVI pure before it comes in contact with a pollutant? I have some ideas, the beginnings of a plan. We need to be ready to submit a bid the beginning of next year. That's six months. Figure another six months till the military makes up its mind. So, a year from now, we're in the remediation business."

Marta and Jim looked at each other and shrugged. Eva had said the magic word: plan.

"Wait a minute," said Marta. "You may be superwoman, but developing a new technology, creating a manufacturing plan, a logistics plan, a cost accounting system for the project, *and* pulling together a comprehensive proposal in six months? That's impossible! You could work around the clock for six months and you still won't be on schedule."

"That may be true for other people," said Eva, "but I can do it."

Jim said, "Do we have to do this one? I mean, there's no harm in bidding on the next project. Lord knows there's enough pollution to go around."

"No." Eva's voice was emphatic. "This is the one I want to start with. I didn't say it would be easy, but I can do it. Nailing this contract would put NMech at the forefront of ecological reconstruction. Granted, there are some problems, but everyone faces the same problems."

"Problems? What sort of problems?" asked Jim.

"Pure iron or ZVI combines with anything it comes into contact with. Mostly it rusts since there's plenty of oxygen in the air. So the biggest hurdle is keeping an inventory of ZVI. Most people fabricate it and haul it to where it's needed. That's expensive. I have a better approach."

"Which is…?"

"I'll get to that. Solutions are simple. It's framing the question that's hard. I can explain once I give you some chemistry background. Let me list the challenges first, then we can talk about the details."

"What's the second problem?"

"Here's where it gets interesting. ZVI is way more effective when it's nano-sized. I'm talking many times more effective."

"How come?" asked Jim.

"Because as the size of the ZVI particles decrease, the proportion of surface atoms increases. Then there are more available atoms craving more interactions with the polluting substances. But that creates a drawback. The nanoparticles are so effective that they consume themselves rapidly. So it's tough to maintain a supply of ZVI. These two problems are matters of logistics, not chemistry. So far, nobody's been able to keep enough ZVI on hand to be effective in a project this size. Remember, we're talking about cleaning an entire century of gunk."

Marta was nodding. She had subvocalized and was peering into a heads-up display. Eva guessed she was accessing data on ZVI.

"Bottom line? None of the remediation companies knows much about nanoscale production. So what if we've never cleaned up an ammunition dump? Nobody else has either. But we know more about nanotechnology than anybody. And I'm telling you, this is going to be big. We get this contract and a relationship with the military and we have a chance to become the biggest company in the world. The military does not write small checks."

"Uh, Eva. Aren't we getting a little ahead of ourselves? We still have to win the contract," Marta said.

Eva grinned. "Don't worry."

"I know, you've got a plan," intoned Marta.

Jim smiled. "Of course she does."

"Yeah, I've got a plan." Eva's grin faded. "And nothing is going to stop us."

"Of course not, Eva," teased Jim. His smile was cut short.

Eva turned to face him. "Let's get something clear," she said. "This is the next step for NMech's evolution. We are going to win this contract. Period. Nobody, nothing, is going to stop us. This is the future. Got that?"

"Sure, Eva," Jim shrugged and backed up a step and offered a mock salute. "No half measures. Aye-aye, Commander. Full steam ahead."

"Can the jokes, Jim. I'm serious."

Jim and Marta sat back in their smart chairs. The temperature in the boardroom seemed to drop. They looked at each other and back to Eva. Marta said, "Eva, lighten up. No one is trying to trivialize your project. You enjoy making money? Fine. You want NMech to be the world's largest corporation? Fine. Let us enjoy our work, too. Let us enjoy your friendship. Look, you and I have come a long way since Harvard. I know I rub you the wrong way sometimes, and God knows that you can push my buttons. But take it easy. Joking can be a good thing, so let's go with the flow, okay?"

"What does that mean, 'go with the flow'?"

"Look at the pictures of the jellyfish on the drapes. They can use the ocean's currents to go where they need to go. Let's not fight the currents. That's what I mean. You can be yourself—determined, intense, and impatient, and that's okay. We're friends. But let us be ourselves, too, and part of that is Jim's sense of humor. Or what he thinks is a sense of humor."

Eva looked at Marta and nodded. "Friends," she said. "Okay, I get it. Fine. Just don't expect a group hug any time soon." Her partners looked at Eva, trying to gauge her. Did she just attempt to lighten the mood?

Eva turned back to the jellyfish display. Keeping to herself, she saw the transparent hoods swaying in the currents. The Medusa-like tentacles held her attention. Some hung for tens of feet, and each was packed with millions of nematocysts—specialized cells that bulged with venom.

17

 HALCYON DAYS

FROM THE MEMORIES
OF DANA ECCO

Zeus created Aeolus to control the wind. Aeolus calmed the wind and seas for seven days during the winter solstice to allow a certain kingfisher bird to lay her eggs in safety.

The bird that merited the Aeolus's care was his daughter, Alcyone. The unfortunate lass had thrown herself into the ocean when she learned that her husband had drowned at sea. The gods then turned the storm-crossed lovers into kingfishers. I would think that a simple rescue would have done nicely—why not have another ship come along? But the gods have their own sensibilities, and human-to-avian transmogrification it was.

Those seven days of calmed seas came to be known as halcyon days. Take the letter, "H" from '*hals*', Greek for seas, plop it in front of Alcyone, ditch the "e," and you have the word halcyon, a nostalgic reference to the sunny days of youth.

Rockford ended my halcyon days. The winter that followed was severe, even by New England standards. There were no calm days for kingfishers—nor, as it turned out, for petrals, nor thunderbirds.

❀   ❀   ❀

If Alcyone was a kingfisher, then Eva was another seabird, the storm petral, the smallest of the seabirds, with a short, squarish body, and dark plumage. It hovers just above the ocean's surface and appears to walk on water. The metaphor was apt. When my parents considered Eva's remediation project, she seemed to be capable of miracles.

She nearly was. Eva attacked the task of preparing NMech's bid with a scorched-earth vigor that would rival General Sherman's march to the sea. She commanded every resource at NMech's disposal and quite a few that were not, in a frantic attempt to meet the submission deadline for Rockford.

If Eva were a storm petrel, then I was a thunderbird, a truculent and quarrelsome fifteen-year-old, creating storms as I flew. My parents mostly ignored the outbursts and tantrums. They could see me struggle to mature and they remembered their own painful rites of passage through adolescence. Eva, however, was beginning to fear that the bid would not be ready on time, and she lacked the time or the emotional resources to be empathetic with me, or to be patient.

She also lacked a model by which to put my behavior into perspective. A part of her was eternally juvenile, stunted, unable to follow me into adolescence. At another time in her life, she would have accommodated a new dimension in our friendship. But she was possessed of a single focus which brooked no competition for her attention.

She was not the only one of us with tunnel vision. My parents and I were blind to the demands she placed on herself, and the consequences of those demands.

It was a small thing, our spat. How many great events turn on a small detail? That day, I was fueled with bravado that went beyond the scope of our usually playful competition. Someone who understood that teen moods 'blow in, blow up, and blow out', to quote Winston Churchill, would have taken a deep breath, counted to ten, and ignored my bratty manners.

I wish Eva had ignored me. I truly wish my mother had.

There's a saying that if a butterfly alters its path, then the course of history is changed. The Butterfly Effect, some call it. That's a bit too philosophical for me, but my run-in with Eva about butterflies did indeed change history.

Just before I stormed out of Eva's work area, my mother and I had pondered how a butterfly emerges from a cocoon. Her objective that day was to place science within the context of mystery, to find the sublime in nature. Butterflies lack teeth, my mother said, so they couldn't chew their way out of a cocoon. If they were to secrete a caustic substance to dissolve the cocoon, would that not burn their delicate wings? My assignment was to look for the answer in the world of science but to preserve the sense of wonder. Awe and humility are essential research tools, my mother said. Science might have an explanation, but attunement with nature's mysteries hones the researcher's scientific intuition. Seek awe, my mother said, and you'll find science.

I did the opposite. I turned clever. I tried to stump Eva rather than sharing my excitement.

The timing of my display of pride was bad, very bad. Eva

was racing to complete NMech's bid. Her usual short supply of patience was long since exhausted. When I nagged and teased her, she snapped. What she said to me wasn't important, but how I reacted had a lifelong impact on Eva and my family and ultimately, the world: I burst into tears.

My outburst would have blown over as quickly as a summer squall but as I hurried from Eva's lab, embarrassed by my artless attempt to play the bully and stunned by the strength of my reaction, I ran into my mother—literally. We nearly tumbled to the floor. Then chagrin escalated to humiliation. The last person on earth I wanted to see was my mother. She held me and kissed me and wiped my tears with her thumbs, as she had when I was a child. Out of the corner of my eye, I saw a small group of lab techs watching us.

Now my mortification was complete. I screamed at her. Eva heard me and came out of her lab with a look of confusion and concern. I ran out from the work area, out of the building onto Boylston Street, through the Public Gardens and the Commons, running until the tempest passed. The outburst was short-lived but the damage was permanent.

In my meditation, I return to that day to comfort my mother, Eva, and the child Dana. I return not as an older version of myself, not a wiser manifestation of the child, but as something ageless. I wrap my arms around the three figures to hold them intact. Fractures race along fault lines deep within the foundation of each one's character. My strength flows from the present. It is tangible and luminous, like fire from the Sacred Heart of Jesus. My love for these ones fuses and anneals the flaws. The fire gathers into plumes and becomes an archangel's wings, softly drawing gall and malignancy from Eva, and she knows peace. The alar radiance has a quilled sharpness, too, and it lances my mother's greatest fear, that I would inherit her pain. Hot infection spills out of her in pustulant colors and she sighs deeply in relief. Then the child—always blameless—turns transparent and the angers and debts of these two women pass through, unretained.

This fine meditation brings me a moment's relief. But the mighty seraph who returns to that moment to give succor is utterly impotent. My mother had previously sworn an oath. *If she crosses a line that involves Dana, we will not have Eva in* any *of our lives.*

When I ran from Eva's lab into my startled mother's arms, misunderstanding animated her vow. In that moment, her oath, sworn years earlier, was fulfilled.

I never learned what transpired between my mother and Eva after I stormed out but when I returned, a changeling had replaced Eva. The substitute was cool, polite, and distant to me. She would sport no teeth, exude no caustic dissolvant. What emerged from her cocoon was not a monarch or a swallowtail, but something dark, blood red, and fearsome.

18

WHOM THE GODS WOULD DESTROY

Eva worked with the consuming passion of a New World missionary. The technical challenge was simple to describe—keep ZVI immersed in an inert gas like helium until it was injected into the polluted river. Expose ZVI to pollutants and you get remediation. Expose it to oxygen and you get rust.

The business challenge was to prove that NMech could provide adequate supplies of ZVI to keep the operation running smoothly. All of the other bidders relied on off-site ZVI manufacture. Transporting the pure iron to the remediation plant increased their costs and risks. NMech's solution was elegant and unexpected. In theory, it looked simple: combine known elements in a new way.

In practice, it looked impossible. How could NMech produce a working model in time?

Eva feared missing the deadline. *At this rate, I won't make it,* she thought. *I have to speed the process.* She reviewed her notes and considered her progress, and the tasks that remained. The science wasn't an issue. The solution she was developing was based on nanotechnological developments dating back to the early 2000s. She needed neither new technology nor methods in engineering. The scale of the project was the issue. She needed more time.

Eva ran her simulations, as she had a dozen times. She changed variables at each step, and then ran the simulations again. And again. The results were maddening and consistent: she would not meet the deadline.

If I could work all twenty-four hours of the day, I could do it. If Marta or even Dana understood the chemistry we could make it. If I had an extra couple months, I could do it. If Jim could write the proposal, even just be here for moral support. She couldn't add hours to the days, or days to the month, and she was working as hard as she could. If only she could think faster and move faster.

Then an idea struck. Eva subvocalized and called up a series of neurobiology texts. It looked feasible. *This is Marta's area,* she thought, *but I'll be damned if I'll let her in on this. She'd have some objection or another. But if I can make this work, I can do it. In fact, this may be even bigger than remediation.*

Eva read more. *There. I* can *do it. I can achieve things that humans only dream of. Then we'll see about Marta Holier-than-Thou. Jim will have to see me for what I am.* She checked the texts one last time and headed to an NMech pharmaceutical laboratory.

⚛   ⚛   ⚛

A quarter century earlier, Eva's older sister, Gergana, and the anti-quarian, Coombs, and an English teacher named Erickson had all urged Eva not to ignore stories and literature. Understand yourself, they had argued, and you will better understand your science. But Eva ignored all three warnings, like Peter's three denials before the cock crowed.

The lessons of literature were lost on Eva. The tale of Bellero-phon or Icarus might have served to warn her before she began her own flight to Mount Olympus or to the sun.

                ❀   ❀   ❀

It worked. Damn, this feels good! Going to do this yet. Look out world, here I come. This project is mine and nothing *is going to stop me.*

And the chorus from the Table of Clamorous Voices was sweet and, for once, harmonious. It sang on and on and on and Eva sang with it.

19

IN DREAMS

BOSTON, MASSACHUSETTS

AUTUMN, 2043

Jim Ecco was jittery. He might as well have chewed a crop of coffee beans. The smart bed could not lull him to sleep. Nor could it dampen his movement enough to protect Marta's fragile slumber.

"Querido, what is it? What's troubling you?" she asked in a strained voice.

"Bad dream."

"Come here," she said, and reached out for her husband.

"I can't lie still. I'm sorry I woke you."

"Querido, come here. Let me hold you and you can tell me about your dream."

Jim sighed. The dream was confusing, upsetting and finally, ludicrous—not one he cared to recount. He closed his eyes and breathed in through his nose, and then exhaled through pursed lips. He repeated the exercise three times. Tonight, the rhythmic cycle of inbreath and outbreath brought no peace.

"Marta, I'm scared." He laid his head in the crook of her left arm. She wrapped herself around him and reached her right hand up and stroked his hair.

"Tell me your dream." She stroked his forehead until she felt him relax a little.

"We were at home. I saw white ash falling from the sky, like something had burned. I didn't know where it was coming from. I went outside to look and the ash burned me where it touched me. I tried to warn you to stay indoors, but you couldn't hear me. I wanted to shout but I couldn't make a sound. You came out to see what was wrong. Then you were burned, too."

"We were afraid that Dana would come out. We saw him at the door and shouted for him to stay inside, but he came out anyway. The ash landed on him, but he wasn't burned at all. Dana just brushed it off and said, "I tried to tell you but you couldn't hear me.""

"Then the dream shifted. Now it was just me. I was in an old-fashioned stationery store, the kind that had antique post-cards. I was looking at different places I might like to visit. When I looked up, I saw superheroes from the graphic novels I used to read. I remember Superman in particular. I can't remember who else. Then I saw my mother. She was angry. She reached over and touched Superman and he turned white, like plaster. She had drained his life force. She came through the store and touched the other superheroes and took their vitality as well. Then she was reaching for me. I was scared. She touched me but nothing happened. I realized that my own superpower was that I have good boundaries. That was my superpower. Weird, huh?"

Jim was quiet for several minutes.

"That's the whole dream?" asked Marta.

"Yeah. It was scary and funny at the end. Weird."

"Well, I think that's a pretty good superpower," Marta chuckled. "Hello, Boundary Man," she said and in a moment, they were both laughing.

"Still restless?" Marta asked.

"I can't sleep," Jim said.

"Come a little closer. You may be Boundary Man but I'm a bohique and I know what's good for you." She rolled him onto his back grabbed his wrists and pinned him on the bed. She straddled him. "Here comes your medicine. A wise woman's orders."

Later, Jim, eyes wide, decided to get up rather than wait for morning. He slipped out of bed and looked at Marta. She had one arm flung up over her head and the other down by her side, as if she were demonstrating the size of the big fish that got away. She looked so peaceful in repose. Maybe her pain was gone for the rest of the night.

He'd left his clothes on a chair. He picked them up quietly and went into the bathroom to dress. He touched the wall to turn on the brightwalls and swept his hand down to keep the light low. He subvocalized and left a message for Marta that he was going to the office. Maybe he could do something useful as long as he was awake. Eva'd been working nonstop and promised results soon. He thought he'd go to the office and see how she was getting on.

The night air was cold. All the science in the world, he thought, and we still can't touch the weather. Maybe it's just as well—we'd just screw it up. Jim subvocalized a command instructing his clothing to warm him. His shirt had an inner layer of silk-like textile embedded with carbon fibers against his skin, and an outer layer, indistinguishable from cashmere. He wore denim jeans and a lining on the inside of them warmed. He tugged the back of his shirt collar and felt a moment's resistance before it relaxed and allowed him to fashion a hood around his head. He invoked a heads-up display and from the transportation options, he selected a P-cab, a driverless personal taxi. He reached a corner parking lot where the car waited for him, glowing to identify itself.

By the time he reached the NMech offices on Boylston Street, he was warm. He left the cab and approached the building. After palming the door for entry, his clothing cooled to comfortable indoor wear.

Jim took the stairs to the sixth floor executive offices. He thought he would review the plans for the Rockford remediation project. He wondered if NMech would be ready to submit a bid on time. Perhaps Eva had managed some kind of breakthrough.

When he reached the executive suite, he was startled to see the entire floor alive with light and color. The brightwalls were dimmed, but so many holographs were illuminated that the suite resembled an outdoor celebration lit by paper lanterns. He saw displays of graphs, flowcharts, architectural drawings, and diagrams that had no meaning to him. He caught movement in his peripheral vision, almost too fast to notice, and he followed the blur to Eva's office. She was coming back out. They were about to collide but Eva stopped faster than he thought possible.

Jim studied her. She was flushed, and a sheen of sweat made her glow. For a moment he thought he'd stumbled into a holographic display. She looked up and smiled. "We're going to do it, Jim." Her voice was uncharacteristically animated, loud even. "We're going to do it. I'll show you."

Eva grabbed his hand and started running towards the conference room. "Slow down, Eva," Jim said. "I can't keep up with you.

What's going on?"

"I forgot. You move slower. I can fix that. First, I show you the proposal."

"The ZVI bid? I didn't think we were going to make it," Jim said. He looked at her more closely. "Eva, what's happened to you? You're running around like a crazy person."

"Ha! You know better than to call me that. But for you, all is forgiven. Come. Look!" She pulled him into the conference room. The glow from a dozen displays was unsettling. They were like grinning Jack-o'-lanterns. She pointed at one, then another and another. "See? See? Is ready. Is ready."

Jim stood still and took in the room, understanding nothing. He looked back to Eva, still clinging to his hand.

"Eva, are you all right?"

"Better than ever, Jimmy Boy. We make proposal."

"Eva, you're acting strange. You're even talking strange. You're starting to scare me a little." Jim tried a smile, to soften his words, but couldn't move his facial muscles out of any arrangement other than slack-jawed astonishment.

"Not strange, Jimmy Boy. Alive." Her words tumbled out in a rush. "They say people use ten percent of potential, but I use more now. Now I deal with all of, of…of what's held me back. I can even deal with you. Come here."

Eva was still holding his hand. She reached up with her other hand, behind his head and grabbed a hank of his hair. She pulled him roughly towards her. He felt an unexpected strength in her grasp. She leaned up and said, "Kiss me."

Jim stiffened.

She repeated, "Kiss me. Isn't that what Marta says to you? Kiss me!"

"Eva, you're freaking me out." He tried to keep his voice steady.

"We're friends. We've always been friends. But I don't *want* to kiss you."

"Yes you do! You hide it all these years."

"No, Eva, I don't. Your friendship means too much to me."

"I've done everything for you, Jim. I kept you out of jail, yes? I helped your wife with her public health, yes? I make you a lot of money. I teach Dana the things that Marta couldn't. I even help you get married. Now it's time for you and me. Now I'm going to take care of you." She pulled him down again and smashed her lips against his. Jim grasped both her wrists in his hands. He held her at arms' length.

"Eva, this is not what I want. I think you've been working too hard." He saw her face turn slack with shock. "Please, Eva, I care for you as much as anyone in the world. Anyone. *But you are my friend, and I don't want to lose my friend.*"

Eva twisted and struggled to free her wrists. Jim gently but firmly pushed her away and said, "Eva, I don't know what's happened to you, but you're not acting normal, even for you." He tried to grin. She did not respond. "Listen, I'm going home. I'm not going to mention this to Marta—to anyone. This never happened." He backed up with the same care he might show retreating from an agitated dog.

With a speed that astonished Jim, Eva leapt forward. She reached out and grabbed Jim's wrists. Her grasp was like iron. He was trapped. He looked into Eva's eyes, now twitching, feral, and in a sad and quiet voice, said, "Please, Eva. You're breaking my heart."

She let go and slowly crumpled to the floor. Jim turned back to her, but she held one hand out in a 'stop' gesture. Then she turned away.

Jim left the NMech building, numb to the cold even as his

clothing refashioned itself to provide warmth. He looked up from the street to the sixth floor, the executive offices. The bright lights of Eva's holo displays were flickering out, one by one.

Jim started to walk. The weatherproofing properties in his jacket were fully activated and repelled each of his tears.

❁ ❁ ❁

Eva lay in the conference room. Something held her fast to the floor, something more than anguish, fatigue, or gravity. Her muscles twitched. At first it was a tremor, then a shiver, finally a feverish seizure. Her eyelids spasmed, like a parody of blinking back tears. She tried to subvocalize a message, to recall Jim, to entreat, to apologize. But she could form no words.

Then images replaced language. She watched from infant eyes as Mama and Papa looked at her, first with pride, then with horror. She felt Gergana's arms and listened to her songs, then heard her screams. She saw Bare Chest's face, looming and leering, then paling in death. Doran's fat wattle reddened as he strangled Gergana and then bled as Eva strangled him with a length of piano wire wrapped around wooden handles.

Then blackness.

Slowly, consciousness returned. The holographic displays had extinguished themselves. She opened her mouth to subvocalize, to bring the displays back. She had work to do. As soon as she moved her lips, she heard a terrible cacophony, a roar from the Table of Clamorous Voices that demanded her attention. The loud voices, the soft voices—they were now unregulated by any agency, any construct. Thoughts and memories, images and stored sensations, rushing up from the deepest trenches of her unconscious. She was overwhelmed.

When Eva was an infant, Gergana's presence helped her to manage the growing din of sensory impression. The din became a roar after Gergana's murder and organized into the Table of Clamorous Voices. Eva invested Jim with the role of mediator, regulator of the Table, and the fantasy role of mate. The illusion helped her weather her inner turmoil in order to meet the demands of the saner world around her. But flesh-and-blood Jim Ecco had just destroyed fantasy Jim Ecco, the construct. The mediator was gone.

Eva lost consciousness again. Her body took to repairing the damage inflicted upon it over the last many days. Her swollen and overworked adrenal and pituitary glands relented. Hypopituitarism replaced her chemically-induced hyperpituitarism, fatigue replaced zeal, indifference replaced libido.

Time passed and Eva awoke to disoriented incomprehension. Was it day or night? Had seconds passed, or hours? She had a pounding headache and her vision had diminished to a dark tunnel, like looking through the wrong end of a telescope.

She tried to move, to organize her thoughts. These tasks seemed herculean. She rolled to her desk and pulled herself up. She saw her coffee mug, still half full. With a grimace she swallowed the cold liquid with the bitter ingredient that had permitted her to work as quickly as she had. It wasn't enough.

Her overtaxed endocrine system was in a state of rebellion. It ignored the chemicals she ingested. There had been too many demands and not enough rest. She'd pushed her body past Mother Nature's limits for this wondrous design, this human form. Now she was weak, unable to focus. Her body demanded rest to repair the damage.

I just need forty eight hours. I can sleep when I'm dead, she thought, and mixed another cup of the adulterated beverage. The effort was almost beyond her. Soon she would break a trail into new territory—all propulsion, no rudder, and with an impaired captain at the helm.

❀   ❀   ❀

Ah, that's better. I don't care what it takes. Rockford is mine.

20

DEBATE

A panel winnowed the field of prospective vendors to two finalists: established remediation leader, CleanAct, and upstart NMech. A year before the plant was intended to open, the competitors met to address the bid committee, a debate to help decide a winner.

CleanAct's president, Fritz Reinhart spoke first. The Chinese-educated Texan of German descent was at ease. He knew several of the bid committee members from industry meetings. Two had worked for him in the past. Reinhart was tall and well-groomed, comfortable speaking to an audience. He wore his thin blond hair in a military-style crew cut and kept a well-trimmed moustache that drew attention to a full mouth with generous lips. His mannerisms were prim, almost prissy, but when he spoke, he transformed himself into a folksy cowboy. He wore a bolo tie, cowboy boots, and a western hat and spoke in an exaggerated drawl. He doffed his hat and bowed slightly—Fort Worth meets Frankfurt—when he took the podium.

"The single reason y'all want to accept our bid is that we've done exactly this kind of work for years. No one has anywhere near the experience we have in remediation." Reinhart paused, making eye contact with each member of the bid committee. He was charismatic and easygoing. The committee leaned forward as one.

"We completed 45 major cleanups in the last five years. Clean-Act's performance exceeded the contract specifications. We were right on time and right on budget. We have six more projects and all of 'em are even a mite ahead of schedule. And we aim to finish ahead of schedule on this one, too. That's our corporate style. It's also a guarantee to you. I promise to this bid committee, right now, that your remediation plant will be fully operational three weeks before the end of the performance clause in the contract. That's part of our culture: better and faster."

One member of the bid committee broke in with a choreographed question, a softball objection intended to appear challenging. "But the bid requires that you use nanoscale ZVI. You have no experience with nano production. And now you're promising to finish early? How are you going to make that work?"

"Now that's a good question. Heart of the matter, yes sir."

"Yes, Dr. Reinhart, it is the central issue. How can you ensure that you'll have enough of the ZVI in nano form? And how will you keep it safe? After all, you have no experience with it. Mismanagement of nanoscale materials can be hazardous."

Dr. Reinhart drew a handkerchief from his inside breast pocket and mopped his forehead. He rubbed his chin. He might have appeared flummoxed by the question but his confidence never wavered. "If y'all are worried about hazards, I'd look to that river there. That's what's hazardous and we aim to clean it. As far as safety, well, we have an effective approach. We'll flood the ZVI storage building with pressurized helium—good, safe, inert helium—before one particle of ZVI goes down the hatch. If even a single atom of helium escapes, we'll know. We don't expect any leaks, no sir, none at all, but if there are, we'll find 'em and fix 'em and *still* be on time and budget. From transport to operations, the ZVI stays in helium so it doesn't combine with anything at all until we inject it into the river."

"But you have no experience with ZVI." The friendly inquisitor pressed for more.

"True. But we have ourselves a real simple solution. We bought the experience."

The Committee, dutiful and attentive, chuckled.

"I'm pleased to announce that CleanAct has acquired FeFree, the very best producer of ZVI. 'Fe' is the chemical symbol for iron, and we think FeFree has the best ZVI fabrication process in the world. We don't have the experience to create the stores of ZVI that y'all need, but FeFree does. So, we bought 'em, lock, stock, and containment chamber. Problem solved.

"So, ladies and gentlemen, CleanAct's approach might not be sexy, but it works. Now, let's take a peek at what NMech proposes. Those Boston folks claim that they can convert carbon atoms into iron atoms to solve the logistics problem." He stared for a moment at Eva Rozen and then started to clap. "I have to give you a hand, Dr. Rozen. Rewritin' the laws of physics. Now that's one darned good trick."

He failed to see the tightening around Eva's eyes, the bunching of the muscles in her shoulders. Nor did he notice a trembling in her hands and feet.

Reinhart turned back to the bid committee and pressed on. "Now, I'm not the brains of our outfit. I just give our people a little nudge here and there to help keep things runnin' smoothly. But we've got some darned smart folks in Texas. One or two of 'em even went to college in Boston, at Harvard, same as Dr. Rozen. They tell me that you *can* change one element into another, but only with highly radioactive elements. Give 'em a shake and they shed a few electrons. That turns 'em into some other mighty radioactive elements."

Eva looked up. Had the bid committe caught it? Had anyone? No! Her head shook imperceptibly in disbelief. Stupid cows, they were, every single one of them.

Reinhart continued. "Carbon? Can it shed some electrons to become iron? Last I checked, carbon has six electrons and iron has 26. So, carbon doesn't have enough atomic bits to shed. You would need atomic fusion to make it work, mashing your atoms together." He mimed making a snowball, in case the idea was difficult to follow. "You find atomic fusion in thermonuclear weapons. I think we're in the business of cleaning up after weapons, not makin' new ones."

Now Eva grinned.

"Well. Maybe Dr. Rozen wrote some new laws of physics. Maybe it's different in Boston. But in Rockford, we go by the same God-given laws of nature that have run the universe for about four billion years. Give or take a few million."

He winked. Pure charm.

"Mind you, I had our scientists look at changin' atoms all around. After all, if NMech has something novel, it ought to be repeatable. NMech says it's got the experts to do it but we don't see anything published to show just how. Maybe they're just keepin' it a secret, or maybe they're playin' for time.

"Ladies and gentlemen, don't sell CleanAct short in the area of fabrication. We bought the best in the business and we're ready to start—and to finish ahead of schedule. That's our corporate culture: better and faster.

"Let me close by quotin' a proverb from the Bible, 'There is a time to every purpose'—I believe that our time is now and our purpose is to clean up that dreadful river."

Reinhart sat down to applause. He nodded to his rival as Eva mounted the lectern to address the committee. She showed neither embarrassment nor amusement by Reinhart's barbs. She was not a compelling presenter. She tended to speak in a monotone and often employed a technical vocabulary that estranged her audience. Today she started in better form.

"Ladies and gentlemen, thank you for inviting NMech here today. I'm not here to tell you jokes. I don't have Dr. Reinhart's sense of humor. In fact, most people say I don't have any sense of humor at all."

The committee smiled. A good sign.

"Besides, the problem is too serious for quips. Let's start by being accurate. Dr. Reinhart, your Bible quote is not from Proverbs, but Ecclesiastes, Chapter 3, verse 1. This is the smallest of Dr. Reinhart's inaccuracies. Second, Harvard is in Cambridge, not Boston. Small points, you might say. But they reflect Dr. Reinhart's consistent fuzzy thinking."

"His larger mistakes are astonishing in their stupidity. Nuclear fusion is the result of combining the nuclei of atoms, not by adding or shedding electrons. Atoms give up electrons in the normal course of forming molecules. For example, sodium sheds an electron when it binds with chlorine. Is the result dangerous? Radioactive? No. The result is table salt."

"I'm not sure why this committee would entrust the largest

remediation project in history to a company run by a man who does not understand the fundamentals of chemistry. How can this man expect to manage cutting-edge nanotechnology? That's like asking an illiterate to read an anatomy text in order to perform surgery. Reinhart's fundamental ignorance should frighten you.

"As far as their proposal, so what if CleanAct bought FeFree? FeFree is best at producing advertising, not ZVI. If they had a workable solution, they could have made a fortune licensing the process to remediation companies instead of selling themselves to CleanAct. We estimate that the successful remediation of the Nuovo River will use approximately $11 billion of ZVI over the next decade, but CleanAct paid less than that to buy FeFree. Why would FeFree sell themselves so cheaply if their process were dependable?

"But let's assume for a moment that FeFree really can produce the ZVI needed. CleanAct's approach relies on transporting ZVI to Rockford. The problem is safety. If there's a leak in transport, or in the containment module, the helium escapes. Dr. Reinhart, the last time I checked, helium is lighter than air and iron is heavier. That means that if there's a breach, your helium goes north and your ZVI goes south. If you're lucky, it rusts. If not, then it explodes."

Eva saw confusion on the faces of the bid committee and explained. "If the ZVI leaks anywhere from production at FeFree, to transport, to loading into the containment module, you get a cloud of nanoparticles. If you suspend small particles in air, then you risk an explosion. Ask any farmer about the dangers of a grain dust explosion. Ask a baker about flour explosions. There were over a hundred of these disasters in the last century. Talk to the survivors of the Washburn explosion. A grain elevator there blew and the blast leveled two mills and most of the town. Never mind that CleanAct's approach is unproven: it's dangerous.

"Ladies and gentlemen, we will use simple, proven

nanotechnology to fabricate and to safely handle more than enough ZVI. Creating new atoms is not fantasy. Reinhart ignores a half century of atomic manipulation. Go back fifty years. Scientists at one corporation took 35 xenon atoms and picked them up and set them down to spell the name of their company. Wrangling atoms is easier for a real scientist than wrangling cattle is for a real cowboy. By the way, Dr. Reinhart, those are some nice boots you're wearing. You've driven a lot of steer in your time, have you?"

Eva had the committee's attention. "Cowboy Fritz here says that there's nothing published? Perhaps Dr. Reinhart's team should forget about chasing cows and catch up on their professional reading. Scientists started fabricating what are called superatoms in the early 2000s—and they published their work. Superatoms are made of several atoms linked together to act as another atom. If you vaporize carbon and condense the vapor, you can build an iron superatom. It isn't easy, but if Dr. Reinhart were capable of understanding the science he wouldn't stand here and make a folksy fool of himself."

Eva was lit by her own passion—and something more. Her face was pepper red and her upper lip was beaded with sweat. Her movements were jerky and her voice was too loud. The bid committee looked on in growing discomfort. No one nodded agreement. Perfunctory applause accompanied her to her seat.

The committee chairperson rose and thanked the speakers and promised a careful deliberation and a decision once both proposals were reviewed. In truth, the outcome had been decided months ago when a cabal of CleanAct's executives, all ex-military or Department of Defense veterans, sat down with the military command at the munitions plant and hammered out a deal. Yes, there would be competition. The law required it. And after the bid committee's careful consideration of both bids, CleanAct would win the contract, fair and square. It had been decided.

Dr. Reinhart stood and approached Eva with a smile and an outstretched hand. "Dr. Rozen, that was one interesting presentation. I must say, you're a formidable competitor. "

And you're a dead man walking, Eva thought as she ignored the proffered hand and walked past him to join her ashen-faced colleagues. They saw what Eva could not see as she walked back to her seat: the smug grins on the faces of the CleanAct executives.

NMech had lost the bid. In truth, they never had a chance. They had failed to consider the political factors that would guide the selection of a vendor, and moved as fatted calves into a den of hungry bureaucratic wolves.

21

DISASTER

One year later, Eva Rozen, Marta Cruz, Jim Ecco and Dana Ecco gathered in an NMech conference room and watched scores of Rockford's residents join the munitions plant officials and the CleanAct's executives gather for the remediation plant opening. The launch of a dump site seldom generated much public excitement, but CleanAct's public relations department had had a year to feed the public's imagination. A video stream broadcast the event and fed the dreams of delegates of other toxic sites who watched in anticipation. Schoolteachers used the occasion to illustrate the principles of environmental responsibility and scientific achievement. Financiers calculated whether remediation would be the

Next Big Thing. Even the viewers at NMech were mesmerized by the scope of the celebration.

Jim commented on the morbid curiosity that drove him to watch the ribbon-cutting ceremony. Marta spoke of her interest in seeing the river cleaned. Eva offered no motive for watching her competitor's success.

Dana's schooling brought him to the conference room. The remediation project provided lessons in chemistry, political science, history, biology, and social science. His gaze alternated between the vid projections, his own heads-up display and sidelong glances at Eva. At one point, he subvocalized a command to his datasleeve and sent a databurst to Eva's sleeve. She looked over at the boy with a wistful gaze and mouthed, "Later." Dana frowned and turned back to the news coverage of the plant's opening.

The winter morning was unusually warm in Rockford, and the sun shone as bright as the town's hopes for a clean river and economic prosperity. Rings of chairs perched on a temporary stage. The front row was reserved for plant and town officials. Dr. Reinhart mingled with them. He shook hands, slapped backs, and doled out humble thanks and earnest congratulations in equal measure. The Rockford High School marching band entertained the assembled guests from just beyond the stage. Their costumes shimmered, first in blue, then gold—Rockford's colors—and sunlight glinted off their instruments and the decorations on their costumes like the tips of a crackling fire.

The officials and honored guests, the townspeople of Rockford, and viewers around the world all focused on the ZVI containment building, with its inverted funnel-shaped dome that came to be an icon for the project. CleanAct's building had passed inspection after inspection. Experts considered earthquakes, lightning strikes, fires, terrorist attacks, and tsunamis, never mind that the munitions plant was nearly two hundred miles inland.

Their conclusion? The building was safe. This pronouncement was all the more laudable because CleanAct was ready three weeks ahead of schedule, as Dr. Reinhart had boasted it would be. In an age of complex projects with near-zero tolerance for error, most manufacturers were hard-pressed to meet a deadline, let alone beat one. But Dr. Reinhart had made this a point of pride. He'd show those eggheads in Boston a thing or two. "Here in Texas," he'd said repeatedly, "we don't always have fifty-dollar words for workin' hard. It may not be the easy way, but it's the Texas way."

The final safety check had been three weeks earlier. Quality engineers flooded the containment chamber with pressurized helium. Had any of the helium escaped, it would have been detected and the project halted until the integrity of the chamber could be guaranteed. Once CleanAct demonstrated the safety of the chamber, all of the helium used for the test was evacuated through a vent high up on the building and rose safely into the atmosphere. Only enough to surround the ZVI remained.

The moment came to bring the plant online. Wielding an oversized pair of ceremonial scissors, the plant manager cut a foot-wide blue-and-gold ribbon, and then Rockford's mayor threw a ceremonial switch. The plant had actually been brought online hours earlier, again thanks to CleanAct's deadline-beating push to complete the project. All that remained was for the containment building to release ZVI into a production vault where thousands of microscopic jets would spray fine mists of ZVI into a collection tank through which the Rockford Munitions Plant's effluvia now passed.

The process was completely automated. CleanAct's proprietary process assembled the analysis of incoming waste, the moment-by-moment configuration of the ZVI spray heads, the analysis of the output, and the scooping up of the heavy, ZVI-bonded pollutants.

There were backups to the process, and backups to the backups. A redundant operating system ensured that none of the operating instructions became corrupted. If quality control sensors noted any irregularity in the operating commands, the plant would switch to the backup operating system and the cleanup would carry on without a hiccup. It was foolproof, CleanAct said, and the plant officials dutifully agreed.

Inside the containment chamber and unseen to the crowd, the backup operating system was monitoring the containment chamber, as expected. But it was not synchronized to the three-weeks-ahead-of-schedule timeline that CleanAct had as the centerpiece of its winning bid. The backup operating system believed that the containment chamber hull integrity test was to be conducted today. This set of instructions should have been deleted after the successful test three weeks earlier. Was it carelessness that allowed the code to remain? In its rush to beat the clock, did CleanAct miss a crucial step? Or was it an act of sabotage?

Whatever the reason, the backup software overrode the primary instructions for the operations protocols. The redundant instructions ordered external sensors to test for escaped helium as it had been programmed to do. Noting no leaks, the backup operating system concluded that the pressure test was successful.

The next step in the testing process was to purge inert helium. But the chamber now contained ZVI, not helium, and tons of the volatile particles poured out of the purge vent. The heavy iron cascaded down the rear of the round containment dome and into the oxygen-rich air that sustained the lives of the observers. The ZVI was little different from grain dust or flour in its explosive potential. In fact, given the size of the nanoclusters, it was more hazardous by several orders of magnitude.

One spark, source unknown, triggered the blast that incinerated the officials and the guests on stage, rattled windows for eleven miles, and prompted seismologists to report an earthquake at the small Virginia town.

The blast was magnificent, as explosions go. Had the guests been able to describe the last moment of their lives, they would first have noted a powerful shockwave and compared it to being tackled by a steroid-soaked team of football players. Their hands would have tried to clap at the sides of their heads when, a few milliseconds into the event, their eardrums flexed and ruptured. An overpowering flood of nausea would have swept them as their internal organs began to liquefy. Given the ability to continue their observations, they would have noted a strange fog appear and disappear as the air's moisture precipitated and then vaporized in the emerging fireball.

The observers' reports would have terminated as the air flashed to over four thousand degrees. Then the firestorm incinerated the plant and any evidence of the cause of the conflagration.

The blast was loud. A mid-twentieth century battleship's 16-inch guns generated 215 decibels and the sound wave flattened nearby seas. The Saturn V rocket that carried its human payload to the moon generated a decibel reading of 220—five times louder than the battleship on the logarithmic decibel scale. The rocket's sound was loud enough to melt concrete. The Rockford blast was estimated at 230 decibels—ten times louder than Saturn V.

The fireball consumed most of the iron nanodots. There were no structures within the blast radius and once the shockwave passed, the event appeared over. Military personnel from the munitions plant were deployed and they fanned out, tending to the wounded and dazed survivors, pulling bodies from the containment building's rubble.

Those who had been watching the video feed were stunned. Schoolchildren wailed. Financiers winced, seeing an investment opportunity literally go up in smoke. Representatives of other toxic sites cradled their heads in their hands and wondered what they would do next.

A different scene unfolded in the boardroom at NMech. The Cruz-Ecco family stared in horror. Eva Rozen was building models of iron atoms with children's construction toys. She'd create one, and then take it apart and build it again. Eva glanced at the video feed, blank since the explosion, and grunted, "I told them it was dangerous. Maybe they'll listen to me now."

Marta stared at her, a puzzled look on her face. "You don't seem very surprised by the explosion," she said in a casual tone, almost nonchalant.

"Nope. Bad science leads to bad results. I warned the bid committee, but they were already in Reinhart's back pocket. Serves them right." Now she was building carbon atoms. Her hands moved faster than a blackjack dealer at a high-stakes table.

"Do you really mean that?" asked Marta. "The explosion serves them right? Being incinerated is justice?" Her tone stayed gentle, casual and interested.

Eva grinned and ignored the question. She looked at the empty video feed and said, "Well, I guess we're back in business. I don't see any obstacles left. We're a year or so behind where I thought we'd be, but that business is going to be ours."

Marta said, "Eva, I'm a little concerned. You're not surprised. You talk like this tragedy serves some kind of higher purpose. Ever since you decided we should go into remediation, you've acted like winning this bid was a life-or-death matter for NMech. I have to ask, did you have anything to do with this disaster?" Her voice was restrained but her gaze was direct.

"Don't be an idiot. I told them it was a stupid idea. Are you suggesting that this explosion was anything but Reinhart's folly?"

Jim interrupted. "Wait. Something's happening at Rockford. Look."

The video feed resumed as new vidbots came online. People staggered drunkenly, their skin turning cyanotic. Those who had been untouched by the fireball had counted their blessings too quickly. They had breathed a sigh of relief—and inhaled ZVI. Most of the particles had oxidized on contact with air and posed no health risk. Just enough ZVI, however, stayed reactive and entered the onlookers' respiratory systems. The nanodots bonded with the oxygen in the bloodstreams of those rushing to the site of the blast. The iron rusted; the townspeople asphyxiated.

Eva looked at the video feed and shook her head. "Bad science," was all she said.

22

DIAMONDS AND DUST

FROM THE MEMORIES

OF DANA ECCO

Nothing is more compelling than a disaster that's viewed from a comfortable armchair or a barstool, or from miles away in a sixth-floor boardroom. The explosion held the public's attention as securely as an inchworm on hot tar. It was news, entertainment, and a cautionary tale. A cloud of dragonflies—video cameras the size of an insect—caught the explosion's aftermath. Datastream providers quickly packaged a four-minute story arc that began with Dr. Reinhart's polished remarks, highlighted the fireball, and concluded with the grisly asphyxiation of the thirty or so observers who rushed forward after the explosion and inhaled active ZVI particles.

While emergency crews mobilized and rushed to Rockford,

my mother slumped in a smartchair, her head in her arms, resting on the polished cherry wood conference table. Dust motes caught my eye as they twinkled in the sunlight streaming through the floor-to-ceiling windows. The pattern in the drapes was still, as if in respect for the tragedy 700 miles south.

My mother pushed herself up from her chair and embraced me.

My father stared at Eva's retreating form.

She stopped in the doorway, turned and gave a shrug. She ignored my mother and grinned at my father. "Time to dust off our proposal," she said.

"What did you do, Eva?" my mother asked. Her voice was sad.

"What did I do? I warned them that this could happen. That's what I did. I built a better plan than CleanAct, that's what I did."

"I mean the explosion. Did NMech have anything to do with that?"

Eva's voice took on a flat, affectless quality, the studied neutrality of anger. "Marta, you asked me that three times and I'll answer you just once more: I had nothing to do with it. I warned them this could happen. You're upset. Okay—it's upsetting. But you accuse me? Better take a mood block before you say anything you'll regret."

"Eva, there's going to be inquiries, people will look at NMech—"

"Disregard that. NMech is clean. And Marta? We've been friends for a long time. We might still be friends—I don't know, since you contrive to keep Dana away from me. My only advise to you is this: don't push me." She stalked out of the boardroom.

My mother looked at me involuntarily. Her face told me all I needed to know about Eva's accusation. But now wasn't the time to discuss my mother's interference with my relationship with Eva. Besides, I'd figured it out months ago.

I decided to investigate. Maybe I could show that Eva was

innocent. Maybe we could act like friends again. I scanned Eva's datapillars. There was no trace of a databurst transmission that might have triggered the explosion. But NMech was one of the largest companies in the world, and I couldn't scan every pillar Eva might have used. Besides, I just didn't want to believe that she'd have murdered dozens of innocent people just to get Reinhart or the bid committee. On the other hand, she'd been acting strangely in the days leading up to the bid submission, and her behavior had never returned to what is normal for her. I wondered, would she have sabotaged CleanAct's plant?

"Eva's a good scientist and she's been a friend to us," my mother said to no one in particular. She was starting to sob, gulping in big draughts of air, shoulders shaking. "But she's been under such a strain. I shouldn't have questioned her. I have no proof, no evidence other than her reaction, and that's not evidence at all."

My father spoke sharply. "Are you saying she gets a break because she's been under a strain? If she did this, that is?" I think he had already made up his mind. He'd been cool to Eva since NMech submitted its bid and he seemed disinclined to give her the benefit of any doubt.

"I'm not saying that," my mother replied, struggling for composure. "But I'm a doctor, not a judge. Doctors heal sinners and saints. If there's a chance for Eva, we have to help her."

"She gets a pass if she's nuts?" I watched his anger grow. His body stiffened as his muscles tensed. I wondered if this was how it was before he learned to control his temper.

"Jim, would you please listen to me? All I mean is that I'm a bohique. I heal, not punish. If Eva broke the law, then Eva pays the price. But that's up to law enforcement, not me. If there's some way to make sense of this, I'd sure like to know."

My mother and father turned away from each other and lapsed

into stony silence. Everything was upside-down. Normally, my father would be defending Eva, not my mother.

I retreated into my own thoughts. Nothing moved in the boardroom except the dust motes. I watched them, drifting lazily about the room. Something about them held my attention more than the moment-by-moment vid coverage at Rockford. The way they twinkled reminded me of tiny diamonds. The way they moved reminded me of an avian flock. They seemed to move with purpose.

They were Eva's eyes—surveillance motes—tinier cousins of the miniature video cameras that the datastreams used. Each of the motes was a half-micron in size—about 300 times smaller than a human hair. Individually, each possessed the ability to process only a minute piece of information. Collectively, they captured our every word, gesture, and expression.

Eva was studying us from her office, deciding what she would do.

PART THREE

THE GREAT WASHOUT

NEMO ME IMPUNE LACESSIT

'TOUCH ME NOT WITHOUT HURT' OR

'NO ONE PROVOKES ME WITH IMPUNITY'

—Motto of the Order of the Thistle
—See also, "A Cask of Amontillado"
Edgar Allen Poe

PROLOGUE

PUBLIC WORKS

CAMEROON, SRI LANKA, VENEZUELA

2045

STAFF SERGEANT MIKE IMFELD, NEAR WAZA NATIONAL PARK, CAMEROON

"Okay, gentlemen and ladies, listen up. I've got a mission briefing for you. We deploy at 0500 hours, so check your kit before you rack out."

The grumbling started and subsided at once. Mission orders on short notice were the rule rather than the exception. The United Nations EcoForce squad consisted of a dozen infantry troops. In the past twenty-four hours, they'd been assembled in Rotterdam, briefed, equipped, and dropped into equatorial Africa.

In combat boots, Sergeant "Big Mike" Imfeld stood five feet, four inches. He was married to the military—no wife, no children. The chain of command was his family, the barracks was his home.

"What are we doing here, Sarge?" a voice called out from the small assembly.

Imfeld's troops operated with an easy camaraderie. Although the men and women scrupulously observed rank, they considered themselves equals in combat. Imfeld maintained razor-sharp discipline, but fostered an esprit de corps that allowed for informality in the question-and-answer session during the briefing.

"Gentlemen and ladies, we've got a reconnaissance mission to observe a pirate army that's looking to take over a natural treasure. Estimates put the pirates at battalion size—five hundred or so ragtag child-soldiers under the command of a sixteen-year-old leader who calls himself General Ade Aluwa. Don't underestimate this boy. Alexander the Great wasn't much older when he conquered most of his world.

"Open up a heads-up display and invoke a map of Africa. I'll give you a bird's-eye view of the area we're going to recon."

The soldiers complied and peered at the African continent.

"Okay, look along the west, right where the continental coastline juts out westward. Cameroon straddles the equator and looks like a sorcerer's cap. Due west, you've got the nation of Nigeria, which is where Aluwa was born and where he recruited his army."

"Any help from the neighbors?" The soldier's clipped speech marked him as South African.

"No. We'll be operating on our own."

"What makes this here Aluwa a threat?" called out a thick Alabama accent.

"Good question. He was orphaned at age ten, courtesy of the local police. At the time, there were about two million homeless children in the country. Aluwa did the math and figured that he could outnumber the police if he could organize the street kids. He started with a fistful of children and picked off policemen, one at a time, using rocks and luck. Within a few months, he had a platoon of three dozen child-soldiers, a taste for fighting, and a small cache of weapons.

"That was six years ago. Now his army is five hundred strong. He wants to occupy a national park in Cameroon, and we're going to scout the young general's operations. An EcoForce battalion will follow us and persuade Aluwa to find other quarters."

"What are we facing?" called out another voice.

"That's what we're tasked to find out. We think he has precision-guided, rifle-fired munitions—50-caliber explosive rounds with GPS chips and guidance. Our job is to find out what other toys the boys plays with.

"Let me tell you what Aluwa doesn't have: smart uniforms. His troops are vulnerable. You make sure your smart wear is running properly. Your uniforms have enhancements that will keep you invisible, armored, and alive. That and my good leadership, of course."

"Hoo-yah!" A dozen voices shouted out in affirmation.

"Gentlemen and ladies, we are going in full stealth mode. This is recon only. Make sure you check sensors, power, electronic control systems, and armor in your uniforms. You'll be safe as long as you follow me and keep your gear in working order. Your shirt-sleeves and pant legs will transform into bandages and splints if you're injured. Run diagnostics on your biomed sensors. You do *not* want those going silent if you need a medic. You *will* stand inspection before we move out and may the good Lord may have mercy on you if your gear isn't perfect, because I will not."

It was a familiar speech. Imfeld would rather drill his soldiers to death than let them suffer even a scratch from the enemy. And so Imfeld took them through their preparations, like a parish priest leading a responsive reading. He might have given this speech fifty times, and his troops could recite the words back verbatim, but none dared ignore a single syllable.

"Gentlemen and ladies. You have magnetic shearing fluid embedded throughout the uniform. It will turn to armor upon impact from bullets, bayonets, or shrapnel. But it does you no good if it's not working. Check the ferrites twice tonight before you hit your racks. Your uniforms include plastic and glass fibers. They will change color to match the environment and provide camouflage. Do not even yawn before your gear is checked. Some of you might like to survive this little expedition and I will personally wring your neck if you don't."

"Hoo-yah!" The troops called again with one voice. They knew from experience that the worst danger they would face was not the enemy, but Imfeld, if their gear wasn't ready.

"Gentlemen and ladies, fall out!" Imfeld shouted.

Imfeld's orders were crystal clear. His troops took the upkeep of their equipment and uniforms seriously. None of the gear was more critical to their survival than the nanoarmor provided by NMech's military products division as part of the company's commitment to environmental protection projects. Dr. Marta Cruz started the program. She would not have known Sergeant Imfeld in particular. But Eva Rozen did, and tracked his movements, feeding the information to her fledgling Cerberus program.

JAGEN CATER, COLUMBO, SRI LANKA

Nine thousand, nine hundred forty miles due east, and at nearly the same latitude as Cameroon's Waza National Forest, Jagen Cater boarded a train at the Colombo Fort Railway Station in the Democratic Socialist Republic of Sri Lanka. He fell, more than sat, into his customary window seat on the left side of the train. He travelled this route repeatedly, and never failed to gaze in wonder at the rugged hill country, waterfalls, misty peaks, and neatly-clipped tea estates as he travelled the eighty-six-mile route to Badulla. From Badulla, he would travel north to a tea plantation in Kandy, near the ancient royal capital.

Today Cater stared straight ahead, exhausted, and noticed none of the scenery. He slipped off his shoes and noticed that his feet were swollen. Leaning back, he closed his eyes as the locomotive pulled its train out of the station, first by straining inches, and then gathering momentum. The morning was young, yet Cater was fatigued. His muscles cramped and twitched, and his dark skin itched. And his feet! *It must be all this travel,* he thought. This was Cater's third trip to Kandy in as many weeks. There were troubles in the fermenting plant and the tea harvest could be lost. MacNeil Tea Brokers, Ltd., employed Cater to ensure a successful harvest, and so Cater was headed to Kandy.

The harvesting and fermenting of tea is a delicate process that combines farming practices unchanged for thousands of years with technology that was no older than Cater's own five decades. The best plants were reserved for the fine plucking. Harvesters searched for plants with silvery-white fuzz and painstakingly picked only those buds, gathering the tiny blossoms in the early morning to ensure the most delicate flavors for which MacNeil was famous. A healthy bush might produce three thousand buds, enough for just one pound of tea.

When the fine plucking reached the factory it was dried. *It must be the baking equipment,* Cater thought. If the equipment were overheating, the entire spring flush could be lost. And Cater, who never failed to watch the scenery with delight, closed his eyes. He leaned his head against the window and drifted in and out of an uneasy sleep as the train glided east.

Cater felt a hand on his shoulder and looked up, confused. The conductor had roused him. He handed his ticket to the attendant, a man with whom Cater had shared this journey for several years.

The conductor looked at his friend with amusement. "Oh, there must be something going on for you, sir. You will be staying in Badulla? Maybe you have a woman there, eh?"

"What are you talking about? I haven't been with a woman since my wife, bless her soul, passed away two years ago."

The conductor pointed to the ticket, "One way."

Cater stared at the ticket. "Saints preserve me. I've been so tired I cannot think straight. I'll buy another one way for the return."

The conductor peered at his passenger and his voice softened. "Tell you the truth, you don't look so good today. Your face is puffy. Are you all right, old friend?"

Cater heaved a sigh. "Not so good today. I haven't felt this tired since I got the filter to clean my blood."

A thumb-sized packet replaced cumbersome dialysis sessions that required him to sit connected to a machine for several hours each week. Researchers had tried for decades to develop an implantable filter. The problem was that the better the filter worked, the faster it became clogged.

Absolute efficiency in a self-contained module eluded nephrologists until an NMech scientist took a whale-watching boat from Boston to Nantucket and marveled at the giant beasts' feeding habits. The whale's baleen caught his attention. The feeding filters that sieve plankton and small animals from seawater to provide food for the beast reminded him of something. He invoked a heads-up display and studied the anatomy of the baleen. Its elongated pores resembled nephrons, the kidney's filtering cells.

The solution to an implantable kidney filter was slit-shaped pores, modeled on the whale's baleen. The NMech approach permitted a self-contained unit to process variable-sized nutrients and wastes in the blood. Dr. Marta Cruz had been able to insist that NMech provide 10% of its filters to poorer countries as part of NMech's commitment to public health.

Jagen Cater had been one of the first recipients of the NMech IDD—Internal Dialysis Device. It had been a lifesaver for Cater and for thousands of others who suffered from chronic kidney disease. A signal from an NMech medical equipment datapiller kept the IDD operating properly, or else it would quickly become a clogged roll of inert plastic the approximate size and shape of a cigar. The NMech maintenance signal and redundant safeguards were monitored continuously to ensure Cater's survival.

They were also monitored by Eva Rozen's Cerberus program.

NANCY KILEY, PARAGUANÁ WATER PURIFICATION STATION, PARAGUANÁ, VENEZUELA

Five thousand, five hundred twenty-three miles east of Staff Sergeant Mike Imfeld's United Nations EcoForce recon squad, and 9,889 miles east of Jagen Cater's tea estate, Nancy Kiley gritted her teeth and left her smartbed's comfort. Kiley was a good boss. She shared the hardships of her charges, and so she had been taking night shifts, working alongside her subordinates. Just as she fell asleep after a difficult evening, an alarm rang. Cursing, she donned a sun-proof and insect-repellant work suit. Kiley exited her small cabin at El Cerro Rojo—"The Red Hill"—a desalinization plant on Venezuela's Paraguaná Peninsula.

The private cabin was one of the project manager's few perquisites at the desal complex. Scant compensation, she thought, for the time she spent in a landscape slightly less hospitable than the living conditions she imagined a planetologist would find on Mars. Would safety protocols on the Red Planet include thrice-daily examinations of clothing, linens, and shoes for biting creatures? Scorpions, Kaboura flies, and poison dart frogs were among the nasty critters that preferred the cool, dark comfort of Nancy's clothing and shoes. One learned to check, to stay alert.

Mother Nature's a bitch, Kiley thought, not for the first time, scratching at one of the many patches of dry skin that flaked and cracked in the peninsula's arid climate.

Her imprecations were out of character for the cheerful and charismatic woman who inspired loyalty among her staff. She relied on kind words and praise and sprinkled them like the gentle rains that once fell on South American's coastline. But the rains had dried up and her mood had soured. A torrent of maledictions had replaced her upbeat patter. She damned the sun that beat mercilessly on her head, cursed the sand and pebbles that ground under her boots and made walking a chore, and swore at each tormenting species of insects with which she was forced to share the desiccated habitat.

Kiley's compact body cast a short shadow in the midmorning sun. She clenched her square jaw in frustration at what was becoming a hopeless assignment. She squinted despite vision enhancements that included a photosensitive nictating membrane, a third eyelid—biological sunglasses for her ice-blue eyes.

Water, water, everywhere, Nor any drop to drink. Coleridge's couplet was Kiley's mantra. She repeated it while hurrying to the desal plant's command center. *What's today's screw-up?* she wondered as she subvocalized a heads-up display. *I'll be dipped in turtle dung before I have another surprise like yesterday.*

Twenty-four hours earlier, her staff panicked when the nano-controlled biocide levels at the plant dipped unexpectedly. The biocides killed off bacterial contaminants in Cerro Rojo's water. Then, as now, Kiley scrambled out of her slumber to attend to the malfunction. As she watched, the 'cides returned to their normal levels for reasons that were inexplicable. An hour later, when her heart had stopped racing and she was drifting back into sleep, a second emergency summoned her from Morpheus' arms. The desal filters had quit. Electric currents that animated the ion transfer mechanism still flowed, but it was as if the system had stopped listening to its programming. The temporary halt in production scared the bejeezus out of Kiley and her staff. Had it lasted much longer, the 30 million people in the arid cities and hills of northern Venezuela and the island nations of the Caribbean would get thirsty. And thirsty people become angry people—desperate and prone to violence.

Kiley cursed the woman who coaxed her to this hellhole from the security of a government job with National Oceanic and Atmospheric Administration. *I don't know who's the bigger bitch,* Kiley thought, *Mother-freakin'-Nature, or Eva-freakin'-Rozen.*

One year earlier, NMech's Chief Executive Officer linked to Kiley in an abrupt attempt to prise her away from NOAA to become the Chief of Operations for the Cerro Rojo desalinization plant. An installation as complex as Cerro Rojo requires the steady hand of a manager who possesses a thorough understanding of the science behind the miracle.

She had turned Eva Rozen down. "I'm not interested in private industry," Kiley had said, "I do good work here at NOAA."

"You like petty people and petty science?" Eva had shot back.

Kiley shrugged. The gesture was invisible. Rozen waited until Kiley broke the silence.

"Look, I've been in government service for 18 years. Another 7 years and I can retire. NOAA may not be as exciting as your life, but I'll have a nice pension and the time to enjoy it."

"So it's money?"

"Are you kidding? Government work and money do not a partnership make. No, I get to do good science. That's the key, Ms. Rozen—science."

"It's *Dr.* Rozen. Chemistry and computer science. Harvard."

"Well, Dr. Rozen, I have my science and a secure position. Why do I need your little startup?"

"Little startup? Any idea how much NMech is worth?"

"Ms.—excuse me—Dr. Rozen, I don't follow financial reporting. So, no, I don't know. Now, I don't mean to be rude, but I can't see what you can offer that I'd want, and I have some petty science waiting for me."

Eva simply stated a figure before Kiley could end the link. The number was an attention-grabbing number, a round number, a digit followed by zeros, as many zeros as there were digits in Kiley's current salary. When the scientist hesitated, Eva lowered the figure. A second reduction, and Kiley capitulated. In the face of Eva's irresistible force, there were no immovable objects.

A year later, the sugar plum fairies of fame and fortune no longer danced in Kiley's head. This morning's disaster? The desal filters had failed entirely. Not one drop of water was being generated. No one could find an explanation for today's outage, nor determine when the filters might go back on line. This morning, a forlorn Kiley missed her former life in government service, with its scheming competitiveness, its venality—and its boring safety.

From 2,240 miles north, Eva Rozen observed Dr. Kiley's frustrations. She smiled without mirth and entered the results of her observations into the Cerberus program.

23

UNQUIET PHENOMENA

While Marta prayed for the Rockford victims, Eva paced in her office. She was restless. Her forearm itched maddeningly and she scratched it until the skin bled. The rhythmic hiss of her pant legs rubbing together as she walked was a whispering voice that further agitated her. Ssst-ssst. Ssst-ssst. *Too bad! What now? Too bad! What now?*

It was impossible to concentrate. The din from the Table of Clamorous Voices overwhelmed her. The cacophonous shrieks and moans and cries, the accusations and taunts battered her sanity like a tidal wave falling on a seaside village.

Eva struggled to review the past forty-eight hours. An investigation of the explosion was in full swing and NMech's records had been subpoenaed. The company was 'a party of interest' in the search for a cause of the blast. She'd instructed corporate counsel to cooperate. But NMech's many layers of security slowed the company's response which created the appearance that NMech was interfering with the investigation.

What to do about Marta and Jim? They suspected her, too. After the blast, Marta spoke of healing and her role as a bohique—*voodoo doctor is more like it*, Eva thought—but now Marta refused to talk or even make eye contact with Eva. She'd been cool and distant for the last year, blaming her for Dana's growing pains. Marta had seen to it that Dana spent little time around her, a circumstance Eva found surprisingly distressing.

Now this. Was Marta a threat, a neutral, or an irritating ally? Marta's support was unlikely, her opposition uncertain. Jim was a wild card. She had to be careful around him. She thought he was her friend, but he'd spurned her. Granted, she'd been acting a bit out of character, but everyone had ups and downs, didn't they? And the boy, Dana—what could he do? *Plenty*, she thought. *I know how well he jacks and ghosts. He's like a kid brother, but now he's under his mother's control. Damn her interference!* She'd planned to bring Dana into an executive role at NMech. Together, they'd be unstoppable. But now the boy could be the biggest problem.

She shook her head to clear that thought. *I just have to finesse this, keep them all preoccupied until the investigation blows over. Just a couple weeks at the most.*

She continued pacing. Every sound in her office was a chorus of voices, mocking her. *Too bad! What now? Too bad! What now?*

Her desk was as bare as her thoughts, with only a white coffee flask and mug. She nudged these items into place—was that the third time?—to center them precisely along the upper edge of her desk. The walls and carpet were set to a milky white and gave the impression of congealed pabulum.

Coffee. She remembered the coffee. Even though the addition of neurochemicals enabled her to think faster, to work faster, to complete the bid, she'd had no chance. *The damned bid was rigged! Now they all want to blame me for the explosion? Nobody insults me without paying a price.*

She brewed a cup, adding carefully measured drops of the neuroactive concoction, and gulped it down. She was rewarded with a paralyzing stomach cramp and bit back tears. Finally, her heartbeat slowed and became regular. Her skin flushed. She could feel her thoughts reorganize. She was pacing faster now, nearly running. Her arms and legs and hands responded faster than she could ever remember. She felt good.

And the Voices sang in harmony, once again.

What's the plan? She looked around her office. Her eyes fixed on Gergana's brooch framed on one wall, a relic from, well, from before. Her attention shifted to a small terrarium on the credenza across from her desk—plants and flowers from around the world that provided medicines and recreational pharmacopeia for synthesis.

A second planter housed a pair of intertwined green and black vines. An ugly and useless gift from Marta. It was supposed to represent anger and grace, qualities that exist in everyone. Marta and her legends. Yocahu and Juricán. *Give me a break. Who the hell does she think she is, preaching about her gods? She'd be nowhere without me. And what did the gods do for her father, rotting in a Mexican jail?*

Eva paced. Then an idea struck. Would it work? Could it neutralize Marta and Jim? Yes! She calculated the moves and likely outcomes with the cold precision of a chess master. She stared into the terrarium and saw the vines as lambent branches of a flowchart rather than mere plant matter. The divisions and offshoots became the steps she would need to take. It would be straightforward. She'd jacked deep into the legal system before and she could do it again.

Eva subvocalized and as soon as a heads-up display appeared, she mouthed a few commands and found her target. Perfect. The man was accessible. She had to bring him to Boston, unnoticed. The office was too risky, but her home? Yes.

Leaving no tracks, Eva jacked into several secure databases, starting with a United States Department of Agriculture public information portal in Seattle. She ghosted through a half dozen others, soaring on currents of thought, leapfrogging and crisscrossing the country and leaving too many trails to follow.

What if Marta didn't cooperate? That was possible. Oh-so-holy Marta, as if her own foibles made her a saint. *If she doesn't cooperate, then I'll run the company myself and Dana gets a shot at management sooner than I expected. I'll give Jim a second chance. He was a good friend. Maybe things will turn out differently this time…*

Eva paced and scratched. What if the investigation into NMech led to Cerberus? Impossible. No one would find a link between Rockford and Cerberus because there was none. And Cerberus was secure. In the meantime, she would prepare for a special guest. She rubbed at the raw skin on her forearm. The damned itching wouldn't stop. *Never mind, I have work to do, accounts to settle.*

It was time to let Cerberus off his leash. Time to cut off the leeches, starting with the soldier and the scientist and the tea man. Once she was certain of Cerberus, then all the thieves who'd stolen her time and her money would be on their own.

But first, she thought, it's long past time to settle a personal account. It was a small matter, a point of personal pride. *No one insults me with impunity. We'll see who's the runt.*

Time for a reckoning.

24

A GENTLE TOUCH
IS ALL

FROM THE MEMORIES
OF DANA ECCO

The next time I saw Eva Rozen was almost like old times. She was playful in her own way. If she'd just caused a disaster that killed scores of people, then she didn't show it. True, her skin was flushed, her hands trembled, and her eyes had a nervous tic, like she was winking at me over and over, but Eva was always unpredictable. Besides, we'd always gotten along like peas and carrots—until my mother pulled us apart, that is.

I was trying to come to grips with the disaster I'd just seen. None of this had made any sense to me. I was mature for my age, but a fifteen-year-old can draw just so much from experience. Even a fifteen-year-old who's closer to sixteen.

I was alone in the conference room. My parents went home. I told them that I would take a P-cab back to Brookline. I heard the door open behind me. Even with my back to the door, I knew that Eva stood there. Everyone has a unique sonic signature, although hers is more like an absence where there should be a presence, like a chalk outline where a body had been. I sensed a small hole in the air currents that blew into the boardroom when the door opened. She displaced so little air that she might as well not exist.

She stopped just inside the door. Neither of us moved nor spoke for at least a minute. I was looking out the window and reached my hand up over my shoulder to offer an upturned palm, like a back-facing beggar. I held my hand loose, no tension in the fingers, and kept my gaze forward. Eva walked forward and slipped her hand into mine. The skin felt leathery and hot. We held hands for several moments, me with my back to her and Eva erect, looking over my shoulder. She subvocalized and the windows became mirrors. I was suddenly staring into the eyes of—of what? A murderer? A misunderstood friend? All I saw was anguish.

"You scared of me, Little One?" Her voice was uneven but her grip on my hand was as steady as a sailor on a tiller. I squeezed her hand once to signal that I wanted my own hand back. She didn't let go but leaned over me and placed her other hand on my forearm. Her strength seemed to have grown and I was pinned in place.

"You think I did it?" I was still thinking over her first question, whether I was frightened by my mentor and friend. I got the feeling that Eva wanted an answer now. I had no time to ponder the day's events.

"I don't know," I said. "Either you did or you didn't." She made no move to release me. "If you didn't, then there's nothing to be afraid of, right?" She shook her head.

"Answer the question. Do you think I caused the Rockford blast?"

"Eva, I've known you my whole life and I've seen you when you're mad at somebody. Maybe you've even been mad at someone to the point where it was fatal. I don't know. But I've never seen you hurt somebody who didn't make you mad."

"I'm not going to ask you again," she repeated. My hand was beginning to ache in her grip. It was cards-on-the-table time.

"If you did, then you're not the same person I know. So, I would say, no, Eva Rozen didn't cause the explosion."

Not good enough. "So, I'm crazy like Dr. Jekyll and Mr. Hyde? My evil twin did it?"

"Oh, crap, Eva, give me a break." That was better. "I'm not even sixteen years old. What do you want from me? Word games? Then do a crossword puzzle. You want a diagnosis? See a doctor. Now it's Jekyll and Hyde? What, you're suddenly reading fiction?"

"Funny," she snorted. We were heading towards rapprochement.

"Tell me, yes or no. Do you think I blew up the Rockford plant?" I tried to turn around to look at her but her hand remained on my forearm and I remained in place.

"Eva, I don't believe you blew anything up. But—you have been acting weird lately, and other people are going to look at you for Rockford. I hope you've got a good alibi."

There was a sudden rigidity in her bearing. She pressed down ever so slightly on my forearm, relented and pressed again. Just long enough to catch her balance—or download a file to my datasleeve. A gentle touch was Eva's style. She didn't need to touch someone's sleeve to jack it; it was simply part of her own gestural vocabulary. It meant, "Tag—you're it."

"You aim to repossess my hand?" I asked and wiggled my

fingers in her grip. I pitched my voice low and calm. I looked down, submissively. If this was an Eva I didn't know, if this Eva triggered the blast, then I was holding hands with a mass murderer.

She made her decision, gave another snort, and let go. "What would I do with your ugly old hand? Besides, I'm not done with you." She sat down at the table next to me.

"What does that mean?"

"We both know that your mom's been keeping you away from me. But we still have work to do, lessons to learn."

"Don't be mad at my mom, okay? Everybody's mad at everybody else now. She needs to cut you some slack, but you need to be a little more…normal."

"You say so."

I think she intended her voice to be flat, but there was strain in it, pain as well. My left hand moved of its own volition to cup her face. She stroked my hand for a moment. Touching her felt good, despite the odd texture to her skin. I leaned forward with my eyes closed. Our foreheads touched and we sat in silent communion for what could have been just seconds or maybe minutes.

"Little One," she said quietly. "I still have two important lessons for you but we have to be careful. Your mother is—"

"She's afraid of you, is what she is! She doesn't trust her own son!" I nearly shouted. My own emotions were still quite volatile.

"She's concerned, Dana. Concerned. She's wrong. She hurt my feelings—a lot—but she cares about you. You think you're being treated like a child but she's doing better than any mother I've ever known. Even if she's wrong about me."

"You know a lot of mothers, Eva?" I regretted the remark as soon I uttered it. Too late to pull it back.

Eva just looked at me for a long minute. I saw an entire childhood pass across her eyes. I saw hope and longing, disappointment and dispossession. Then anger. "One was enough," she said, "and I know that good mothers make mistakes, but they're always looking after their children. You got one of the best."

"Fat lot of good it's doing us right now."

"Yeah, well get off your high horse, sonny boy. I have a plan."

At that, I grinned. "A certified, grade-A Rozen Plan?"

"Exactly."

She smiled back. Not a grin, not a grimace, but something tender that reached up into her eyes. Suddenly I just wanted her to hold me and make everything all right—like a mother does. I was confused, but beyond caring. She could have kissed me then, not a chaste kiss from a treasured aunt, but full on the lips and I would have been her lover. She could have hobbled me, and I'd have been her pack animal and carried the lifelong burden of grief she'd collected.

"Listen, Dana, listen carefully. This is important. Shit is going to hit the fan. You're going to need help. I could tell you what to do, but you'll learn better by figuring it out. So, I have two more assignments for you and then your schooling with me will be complete. You ace this, and there's no stopping you."

I tried to keep my voice steady. "What assignments?" I asked.

"A puzzle and a treasure hunt."

"I don't get it," I said, but my curiosity was piqued. We were Eva and Dana again, the co-conspirators. Mentor and pupil. Hero and sidekick. Friends.

"Follow me," was all she said.

We walked out of the conference room and down six flights of stairs to NMech's street-level atrium. The wide-open area featured trees growing inside, nourished by full-spectrum lighting that radiated from the brightwalls. The area was littered with sofas, comfortable chairs, and small tables that created sitting groups or spots where someone could rest quietly for a few minutes in some semblance of solitude. Sound strips were built into the floors and walls for private conversations, or so that a person could play music without disturbing others. It was a favorite place for scientists to think, and for workplace romances to flourish—an NMech hotspot for productivity, of one sort or another.

At the far side of the atrium, Eva paused in front of a blank wall and palmed a spot on the wall that was indistinguishable from any other spot. The wall opened inward and led to a set of stairs. When we entered, the brightwalls illuminated automatically and we walked down to a basement and then a sub-basement.

"Wow! This is like in the old, old movies." I was swept up in the spirit of adventure. "We should program the room to look gothic."

"Right."

She touched the brightwall and it illuminated in a nondescript gray, casting a pallid blanket over the room we'd entered. Hardly gothic.

In the corner of the room was a smaller room, maybe six feet by eight feet. The door opened to reveal a small table which held several items: a rolled-up dataslate, a set of old-fashioned wrenches, a pencil, a large, circular magnet, some abrasive cloth, and a square box with a button on it.

"Go on in," she said, gesturing for me to enter first. "Here's your first test," she said. "You have one hour to get out of this cell and not a second more. You get one try only. You can use any one of these items," she said, gesturing to the table. "But here's the catch. You may touch only one of these items. You must use whatever item you've touched in some way. And you get only one try to escape. Let's see how much you've learned."

She asked, "Any questions?" When I hesitated, she said, "Good, because I wouldn't have answered anyway. You're going solo. Give me your datasleeve. Come up to my office within the hour and you'll get it back," she said, and walked away without a backward glance.

After I had handed over my datasleeve, she slammed the door shut. I whirled around, confused. I felt a tiny vibration in my feet when the door slammed home and a thrill of fear. What if she *did* cause Rockford and I'd just placed myself in a cage?

My cell had carbon shielding around the perimeter, and a carbon floor and ceiling. It could be harder than diamonds or as brittle as graphite. Maybe I could kick the door open or just break down one of the walls. But Eva had said that there was one way out and that I was only permitted one try.

I started by inspecting the locking mechanism on the door. I couldn't see anything besides an old-fashioned doorknob. No visible biometrics sensors, no old-fashioned combination keypad. I reached to check how sturdy it was, but pulled my hand back. One try.

I turned to the small table. The dataslate was rolled up. Could I use it to reprogram the door? If it worked, I could. Maybe. But I couldn't tell if it was operational. Heck, I couldn't even tell if it was real. If only I could lift it up and examine it.

The pencil was an ordinary #2, made from old-fashioned wood. I could use it to write the Great American Novel but I had maybe 55 minutes left. Not even enough time for a short story, let alone a novel. I doubted I could create a decent three-line haiku poem in that time. But I could use it to poke at the dataslate and see if that works. Would that be within the rules?

I looked at the doorknob again and ruled out the wrenches. The magnet might work if the doorknob were metal. No dice. The locking assembly appeared to be a non-magnetic material. I couldn't think of any appropriate use for the abrasive cloth. I couldn't file my way out in an hour.

That left the square box with the round button on it. I looked at the device. Nothing on the outside of the box gave me any kind of a clue as to what was inside of the box, nor could I intuit anything about the doorknob and doorjamb. Was this Eva's sense of humor?

I wondered how much time I had. With my datasleeve gone, I was cut off from the rest of the world. My pockets were as empty as my inventory of solutions for escaping from this coop. I didn't think a lot of time had elapsed, but in the isolation of a very small room, it was hard to estimate the passage of time.

If I could solve this problem, Eva and I could continue to collaborate. I had to try.

"When you want to hide something, put it in plain sight" was a maxim that Eva had drilled into me over and over. So, I looked around my little cage for something obvious. Eva, for all of her eccentricities, would never give me a test I couldn't pass, and she always kept her word to me. But would anyone find me if Eva went mad and forgot about this room?

Time was running out. My hands were sweating and my mouth was dry. I had an itch on my back that was driving me crazy. I could think only of the itch. If it were not for her instructions, I'd grab the pencil and use it to scratch my back.

Then I saw the answer. I smiled. An elegant solution, simple and economical, like her software coding. I picked up the pencil. It reached the itchy spot on my back and I scratched. That felt good. I stuck the pencil behind my ear and walked over to the door. I grabbed the doorknob, turned it, and walked out of an unlocked cell.

Five minutes later, I walked into Eva's work area, whistling a happy tune. She looked up and grinned for a second and then pointed me to a chair. She tossed my datasleeve back to me.

"You put some nice security on this," she said. "I couldn't jack it, at least not here and not in the time it took you to stumble out."

"I made a few modifications," I said, trying for nonchalance. "What's next?"

"You solved the riddle. Remember, when you face an impossible challenge, your first move should be to look for the easiest solution. That'll probably be right."

"Now I have a self-study project for you. To finish this last assignment, you'll need to use every bit of the chemistry, nanotechnology and materials science, and physics that you've learned." She got up from her chair and walked around her desk. Perched on the edge, she was about as tall as I was, seated.

"Fact is, it's time for you to fly solo. Take on a role at NMech. So here's what you have to do. Here is a list of 34 senior researchers at NMech, 26 department heads and 18 executives." She held up her arm in a transmit gesture and my sleeve pinged receipt of a file. "Your job is to jack every single one of them. Learn the chemistry or physics or materials science of each one of them by ghosting through their pillars. Learn how they manage their departments by observation and by jacking their diaries. Then link to me and we can carry on our conversations again."

"That's a big job, Eva," I said, with maybe a little complaint or trepidation in my voice.

"And you're a big boy. You're fifteen—"

"Almost sixteen…"

"—and you've been taught the science by your mom and me, and your dad taught you some lessons that will make the assignment a lot easier."

"Why all of this? Why jack 78 people's sleeves?" I asked.

"Keys to the kingdom, kiddo. When you're done, you can run the show if you want."

Great, I thought. *I need a one-man corporate takeover like a bald man needs a comb.* But all I said was, "Gee, Eva. Thanks."

"Dana? Get your ass in gear. I have things to do, places to go. Time for big mischief."

Her voice was starting to quaver again and the sound made me nervous. I had a very bad feeling right then, but nothing I could put my finger on, just a sense of foreboding. Had she really been her old self the past hour or so? Or was she just putting up a good front? Either way, the sound of her voice right then worried me.

Still, I did as Eva asked, and over the course of the next few days I would learn more than I ever wanted to know about the business end of nanotech and 3D manufacturing. I also found a strange piece of art in an unlabeled account. Eva would say nothing about the image. No matter how I pressed her, she refused to discuss it with me.

25

SECOND SKIN

Dr. Colleen Katy Lowell, creator of morphing textiles technology, walked along Boylston Street, near the Public Garden, looking into the window displays of high-end clothing stores. She joined the shoppers who stopped to watch as the garments on display morphed from style to style. Colleen's technology had made its first appearance at the expensive boutiques in the heart of Boston. She grinned and rushed past strollers on the broad sidewalk. *Well, lah-di-dah, lah-di-dah, lah-di-dah. I don't need NMech after all.*

She was exhausted and elated after a successful week of around-the-clock negotiations to secure funding to produce her line of nanocouture. She signed three prominent designers on the promise of venture capital money, and the VCs came on board when Colleen promised the designers.

The week had flown by. Meeting with the money people, then the nanofabbers. After agreements were reached, on came the marketing and distribution experts, and channel sales organizations. The manufacturers were the toughest of a tough lot—the few factory managers who understood fashion also understood that they were a very small group and wanted to charge accordingly. Admins crept in unnoticed with food and beverages and crept back out with the trash that the week's conclaves generated. Samples of fabric, design, and prototypes appeared when required and disappeared when no longer needed. Colleen barely noticed the faces of the bearers of these items. She scarcely remembered breaks for food, changes of clothing, or the odd shower. Sleep? Forget it.

In her triumphant, if dazed, march down Boylston Street, she passed the alphabetically-arranged cross streets—Arlington, Berkeley, Clarendon, Dartmouth—and paused at the public library, the nation's oldest. Two heroic bronze sculptures—female figures representing art and science, one holding an artist's palette, the other an orb—flanked the entrance to the building. She felt as invincible as the bronze heroines. What could stop her now?

Colleen reached her building. It was nestled in Copley Square, a hub of business, learning, and leisure. She sleeved past the security pillar and took an elevator to the eighteenth floor. The door recognized her and opened as she approached. She stopped in the entryway, kicked off her shoes, and stepped onto a thick pile rug. Today it displayed a traditional Moroccan design, an ivory background with brilliant blue diamonds. Colleen adored the soft cushion under her feet.

She was shaking with joy, ecstatic at the fulfillment of a dream. There was more work to do than she could imagine, but right now was a time for a quiet celebration. She had done it.

Colleen crossed her living room, picked up a crystal decanter from a sideboard, and poured two fingers of a Laphroaig Scotch Whisky. She swallowed the smoky liquid, letting the peaty Islay malt warm and relax her.

After a moment's rest, Colleen went to the bathroom to wash her face. She noticed a smudge on her sleeve. No matter. She would activate the garment's cleaning properties while she changed it from a business suit to something casual and comfortable.

Dr. Colleen Katy Lowell's last living act was to subvocalize instructions to her datasleeve to refashion the garment. She chose culottes and a loose-fitting top for freedom of movement. She decided to let her sleeve pick the color from a palette that complemented her light brown hair and fair skin tones. The sleeve displayed a selection of reds and Colleen confirmed the choice. Perfect. Designer Bill Blass had said, "When in doubt, wear red."

Colleen never tired of watching the fabric stretch and pull, reforming itself. She imagined that it was like a second skin, conforming to her figure and mood. She stood still as the jacket lost its pockets. The sleeves shortened and the jacket wove itself from an open front to a pullover. The legs had begun to pull up away from her ankles when her datasleeve processed a string of code that lay hidden in her sleeve's memory.

The tightening across her chest was the first indication that something was wrong. Colleen subvocalized but the jacket continued to constrict. First it was uncomfortable, then painful. The jacket compressed her chest and pinned her arms, an anaconda on its prey. She couldn't breathe. She stumbled into a wave of vertigo and collapsed. Pinpoints of light speckled her vision. She tried to call out—nothing but a hoarse whisper. Then, blackness. Her lifeless body lay cushioned on the soft pile of her treasured rug.

Four minutes later the garment relaxed and followed Colleen's original instruction. It morphed into a loose top and culottes. Medical sensors, briefly deactivated, now triggered a distress beacon. The garment began rhythmic pulses, attempting CPR to revive the inert form. A recording of the event would show a spike in blood pressure followed by asphyxiation from a stress-induced myocardial infarction, a heart attack. It was understandable given her workload, a pity given her age.

The fatal databurst had travelled from satellite to satellite, from pillar to pillar, losing its pedigree. It would never be traced from the dataport on Eva Rozen's Cerberus datapillar.

26

DEPARTURES

Colleen Katy Lowell was interred in a beautiful setting on a dreary day. The memorial service was held in Harvard College's Holden Chapel, one of the oldest college buildings in America. The tiny edifice served as a house of worship in 1744. Later, it became part of the College's medical school. The building's diverse history mirrored Colleen's eclectic talents.

Marta, Jim, and Dana sat in the front row of a small group of mourners. Colleen's mother was a convalescent in a Minnesota nursing home. Her father had passed away and she had no brothers or sisters. A college friend, Rebecca Avery, two programmers from Colleen's small company, and a scattering of others rounded out a scant assembly. Avery spoke briefly, and briefly cheered the mourners by describing Colleen's wild streak as well as her brilliance. One of her design colleagues spoke of Colleen's dedication to beauty. The other was mute with grief.

Jim helped Marta stand to address the assembly. Her skin was fever flushed, her pain obvious. "Colleen was different from anyone I know. I believe that everyone has seeds of anger and of grace—human weaknesses and God-given strengths. The way we balance these forces determines our good days and our bad days. Colleen had her faults but she was without guile. She was unpretentious—just look at how she mingled with corporate executives, runway models, and backroom maintenance staff. The years that she spent pursuing her dream testify to her confidence. She was my friend and I miss her terribly."

Marta led the small assembly along a two-mile procession from the chapel to Mount Auburn Cemetery. They drove in silence. The funerary convoy would process past Cambridge Common, grey and muddy in the late winter gloom. The mourners would be escorted along Memorial Drive, a broad roadway that hugged the Charles River. Elm, linden, hawthorne, and lilac trees stood barren in the winter chill, rigid sentinels honoring Colleen's passage.

Not one of the trees at the cemetery, nor the gardens, nor the ponds, nor the dells salved the bitter ache in Marta's heart, neither did they soothe the fever that burned in her face. She summoned the last of her strength to stand alone over the yawning grave and to watch Colleen's casket feed the hungry earth. When the coffin was in place, Marta took a lilac-hued aster to place on the coffin. Ancients believed that the perfume from an aster drove off evil spirits. "It's too late for that now," Marta said, and dropped the useless flower on Colleen's casket and then turn to accept her husband's arm and comfort.

❁   ❁   ❁

It was unlikely that there was a more uncomfortable person anywhere in New England, perhaps the entire eastern seaboard, than the woman who stood behind Marta Cruz, waiting to speak with the grief-stricken scientist.

She was a bookkeeper at NMech with neither managerial authority nor seniority in the company, having joined the accounting staff only months earlier. She recognized Marta—Dr. Cruz—but had never spoken with her. She knew Colleen Lowell from news vids. She had met Eva Rozen once, and then managed to avoid the CEO. That was an easy task. Denise Warren was, after all, just a bookkeeper.

But I've been a conscientious bookkeeper, she thought. *I like it when things balance.* She believed that she'd been given a gift, a sixth sense that prompted her to dig a bit here and there. Sometimes, when she dug a bit here and there, she found something that Didn't Fit. *Not so much a gift,* Warren thought, *but a curse that's cost me two jobs, and now maybe three.*

Her first disaster came two years ago when she uncovered something that Didn't Fit—a scheme to inflate her employer's sales figures. *My luck, I bring this to my boss and find out he's the one who rigged the charade. He gets promoted. I get fired.* Nine months later her intuition led her to discover an innocent error, but the company's restated financial report forced the business into bankruptcy. Warren's position fell to a cost-cutting program prompted by her findings.

So it was with understandable trepidation that Denise Warren approached Marta Cruz to offer condolences, and to bring her Jeremiah-like intuition to bear on an inexplicable series of entries in the NMech accounts receivable department. The funeral of Dr. Cruz's friend was neither the time nor place to discuss a business matter, but the discrepancies had aroused her curiosity, which led to more discoveries. The irregularities would be a serious issue for the annual audit. But what prompted a now-hypothermic Denise Warren to linger at the funeral of a stranger was a bothersome detail that looked, well, criminal.

But what do I know? I'm just a bookkeeper.

Denise blew on her hands and shifted from one numbed foot to the other. Despite the warmth-preserving fibers in her gloves and socks, her hands and feet seemed about as warm as meat in a butcher's refrigerator. When the rest of the mourners had departed, she approached a weary and equally cold Marta Cruz.

"Dr. Cruz, I'm so sorry for your loss, and, um…" Warren stammered and hesitated. Would this cost her job?

"Thank you," Cruz murmured.

"I'm Denise Warren from accounting. I'm sorry to trouble you at Dr. Lowell's funeral, but I need to tell you something. I know this is a bad time, but—"

Jim Ecco stood and placed himself between the two shivering women. "This *is* a bad time. Why are you here, anyway? Did you know Dr. Lowell?"

Warren's eyes turned down and she felt them well with tears. She had visions of losing yet another job. "I'm sorry. I wouldn't be here if it weren't urgent and I don't know who else to turn to. My boss won't listen to me, but there's a problem that will hurt NMech."

"There's going to be another problem if you don't leave my wife alone."

Marta placed her arm on Jim's and looked at the distraught woman.

"I know that I'm nobody." Warren drew in a breath and then pressed on, "I'm an ant."

Marta started. She looked more carefully at the accountant.

"What did you say?"

"I, uh, I said, no, it's not important."

"Yes, it is. You said, 'I'm an ant…'" Marta's voice trailed off. "Bibijagua…"

Denise looked confused. "N-no. Bookkeeper. I, uh, I've only worked at NMech a little while. But I found a problem and it can't wait." She faltered. It was useless. Why would a scientist care about a bookkeeping problem?

Marta looked at the woman. She was pale with cold, fatigue, and fear. Marta took her arm. "Ms. Warren? Are you as cold as I am? Would you like to join me for a cup of coffee? In fact, I could use something stronger. Maybe a lot stronger. Let's find someplace warm, shall we, dear?"

❊　　❊　　❊

Twenty-five hundred miles southeast, on a small island off the coast of the Mexican resort town of Puerto Vallarta, two guards herded prisoner 14162C from his cell at the Isla Maria Madre Federal Penitentiary. The prisoner coughed and reached for a cigarette. One guard told him to get his things together. He was being released.

Prisoner 14162C did not comprehend the news. He had years left on his sentence, assuming he'd survive that long. He'd managed to make a place for himself in the minimum security facility. But he'd aged, and was weaker than a 56-year-old who had not spent most of his adult years in prison. Still, Isla Maria Madre was warm and blessed with fresh air. In another environment, 14162C would have perished from infectious disease or unrestrained violence.

The guards marched him past the prison's encampments, construction sites, and farming areas. They stopped at the prison commissary where he was allowed to purchase two loaves of bread for his journey. The guards could not or would not tell him where he was going, or why. They herded him into a jeep and travelled to a small airfield. Prisoner 14162C was to be flown to the Mexican mainland, and from there he would be taken into custody by someone else. The guards were expressionless. The prisoner was confused, but excited.

When the small prison plane landed at the Puerto Vallarta International Airport, three security agents met 14162C. The prison guards unlocked the man's shackles. The security agents gave him a change of clothing and slapped a narrow strip of nanofabric on his neck. It looked like a priest's collar. They warned him that if he tried to escape he would be subdued quite painfully. One of the agents touched his datasleeve and the prisoner winced and clutched his neck, where the nanofabric had been placed. "That is just a taste of what you'll get if you even look cross-eyed. Understand?" The prisoner nodded and was herded to a small plane bearing an NMech logo.

Rafael Cruz was headed north, a perplexed but willing guest of Eva Rozen.

27

GUESSING GAMES

Marta Cruz watched Denise Warren stare at the place setting in front of her, glance at Jim, and then quickly look down again. Jim studiously ignored her. Dana gazed at her, fascinated. Denise had a round, open face, freckled, and framed by light brown hair cut in a pageboy bob. Her black slacks were an expensive blend of silk and wool, well-tailored and well-worn. A dark purple jacket with a Nehru collar was buttoned carefully over a black blouse. Marta looked at her eyes. Another day they might sparkle inquisitively, but now Marta saw only grief.

Marta felt protective of this woman she'd met only moments ago. She put her hand on Denise's. "I don't know about you, but I'm still cold. Right now I feel like I might never be warm again. Would you care for a glass of wine? And I hope you won't make me eat alone." She caught the waiter's attention and asked for menus. She turned to Denise and asked, "Do you like red wine or white wine?"

The bookkeeper shrugged. "Whatever you're having is fine."

Marta assumed hostess duties. She pointed to the wine list and ordered a bottle of Stag's Leap Chardonnay and one of Cakebread Cellars Merlot. "Please bring us three," she paused and looked at her son, "no, make that four glasses. And some *apertivos* for the table if you would, please."

Wine, water, and plates of bread materialized and Marta asked, "Red or white?"

"Either one," said Denise.

"Oh, my dear," said Marta, "I'm not sure what you think of me, but mindreading is beyond my capabilities. That's my husband's province. In fact," she turned to Jim, "which wine does Denise prefer?" To her puzzled guest, she explained, "He's good at this, you see."

Jim studied Denise. "Red."

"Good guess, dad," said Dana, "but I think you're wrong."

Jim gave his son a look that said, "Don't start with me."

"I don't quite think I understand," said Denise.

Dana turned to her. "It's like Mom said. Some people think Dad's a mind-reader but he just looks for the tiny gestures people make. He sees things that others don't see. But he's trying to figure out why you're here and he can't. That's making him nervous and he guessed wrong about the wine."

Jim said, "'How sharper than a serpent's tooth it is to have an ungrateful child.'"

"King Lear?" asked Denise.

"Excellent," said Jim. "Somewhere in Act I, if I remember correctly. I don't know Lear, but I think every parent has that quote down pat." He grinned at Denise and her face relaxed. They had found a small common ground.

Marta turned back to Denise. "Ms. Warren, this is a game that my husband and my son play. They call it 'reading' people. Do you mind?"

Denise looked back and forth between Dana and Jim and shrugged. "I…don't know what you mean, but okay."

Marta watched as Dana considered their guest for several seconds. Her pride in him helped to balance her grief. Dana was beginning to develop the features of manhood. His face was chiseled, quite unlike Jim's; he looked more like Rafael, her father. Dana had a hawkish nose and pronounced Adam's apple. The hint of a beard that he was developing added shadow to his face. He was built with broad shoulders, like Rafael, and would grow to about six feet, unlike anyone in Jim's family or in her own. He was a unique individual.

Dana looked Denise over and said, "You're a solitary person, but not always by choice." A slight tension appeared on Denise's forehead. "Ah, gee, I'm sorry, I shouldn't have started there. Mom says to start with the things people like to hear."

"How did you know that?" asked Denise, interested and, for the first time since they'd met, a bit more at ease.

"I'll explain everything in a minute," he continued. "You are more orderly than most people. You got laid off or fired before you came to NMech." Dana paused and watched her reaction. "Twice?" She nodded. "Whistleblower?" She cocked her head and stared at Dana before nodding again.

"You thought about coming to the funeral all night and didn't get much sleep. You made up your mind to come at the last minute. You have a cat—is it named Rex? Mom trusts you and she wants Dad to trust you, too. And he's wrong, you prefer white wine."

Denise stared, openmouthed. Jim smiled and Marta beamed at her son.

"How on earth did you know my cat's name?" Denise asked. "Did you get that from your sleeve? I didn't think that was in my cloud data."

"No, that was a guess," said Dana. Marta watched her son. It was his turn to beam. She knew that demonstrating his skills in front of his father filled Dana with satisfaction.

"You have a few cat hairs on your clothing. They're very curly. Only a Rex cat has hair like that. I took a chance it's a male and that you named him Rex."

"Did my son get it right, Denise?" asked Marta.

"Yes, he did," she said, nodding her head. She smiled at Jim, "You must be proud of him. But Dana, what about the rest? The last-minute decision? Job troubles? All that?"

"I'm sorry if I got too personal right away. But you've got cat hair on your forearm and on the bottom edge of your jacket, and on your slacks where they would meet your jacket if you were sitting down. So, your cat jumped on your lap as you were sitting and you were wearing the jacket at that moment. You seem like a careful person—I mean, you're an accountant, right?—so you would have taken off the jacket before you sat down. Or you would have noticed the cat hair if you weren't in a rush."

"You've got a good eye," Denise said quietly. "Now tell me the rest. This actually makes it easier for me to tell my story."

"Okay. Your clothing is stylish, the edges of your sleeves are frayed. So times are a little tough and that points to job problems. You passed on buying new clothing, but made sure your hair was properly cut. You are conscientious, which is why you came to see Mom at Colleen's funeral, so you didn't lose your job because of anything you did wrong. Maybe it was something you did right that got you in trouble?"

It was clear to Marta from Denise's smile that she enjoyed the boy's attention. *He will be quite a prize for a lucky woman some day. Or a lucky young lady very soon,* the proud mother realized. She felt a momentary pang of—what? Not jealousy, but something akin to it. She felt protective. Dana would find someone to love him. She would have to trust that person to love him as deeply as she did. Could anyone care about him as much as a mother?

Her rumination was interrupted as the waiter came by with a platter of appetizers. Crunchy cod fritters, sweet plump cornmeal fingers, and crescent-shaped turnovers, some filled with lobster, some with beef. Steam floated up from the platter and carried a piquant aroma of pepper, oregano, and garlic. The four diners attacked their food. The only sound from the table was the clink of silverware and expressions of enjoyment.

When the waiter returned, Marta asked Denise, "Do you mind if I order for the table?" Denise nodded and Marta spoke for a few minutes in the rapid, guttural Spanish characteristic of Puerto Rico. The waiter smiled his approval and returned to the kitchen.

"This restaurant has the most authentic Borinquen food you'll find in Boston. I've never been disappointed," said Marta.

"Borinquen?" asked Denise.

"The Taíno word for Puerto Rican," Marta explained.

"Taíno?"

"Ah. The indigenous people of Puerto Rico were the Taíno Indians."

"Well, this will be something new for me. It's hard to find any cuisine in Boston other than Italian. Or seafood—but it'll probably be in marinara sauce," said Denise. The family facing her chuckled.

A tureen of black bean soup appeared, following the appetizers. Marta smiled. "Some people say that the black bean soup is Cuban in origin, but I do not accept that. One hundred percent Puerto Rican *puro.*"

They finished their soup and awarded plaudits to Marta for her choices. Then the table grew quiet.

"Suppose you tell me what's troubling you," Marta said to Denise. "Relax, take your time."

Denise Warren drew in a deep breath and exhaled. She lost her hesitant manner. "Okay, here goes. NMech's bookkeeping for accounts receivable—the money that customers owe us—is easy to automate. Same transactions, over and over. Every month the same prescription or the same lease payment for an environmental project. That's the key. The transactions are repetitive, and no one really has to look at them."

Denise continued, a professional in her element. She had the table's full attention. Waitstaff cleared plates, poured wine and water, and left, unnoticed.

"If the accounting system is up to snuff, then you can trust the results, as long as people use the system." She looked around to make sure the family was following her explanation.

"Okay. One more technical bit, then it'll be clear. There are millions of transactions. Accountants, auditors, regulators—they can't check each one. So the auditors pick a sample and test. If there are any discrepancies in the sample, then there's a problem."

Heads nodded around the table.

"Well, I'm new at NMech. I wanted to learn more about my job, so I spent some time looking into the operations. And that's when I found it." The forlorn look returned to her face.

"And *it* is…?" Marta prompted.

"There's, um, too much money. I know that sounds crazy. But revenue exceeds what we were owed. The amount of money that people pay us should equal the amount of money that they owe us, right? I mean, nobody pays extra. The difference was barely enough to notice. A few dollars. Even auditors disregard this small of a discrepancy. But I was curious."

"What I found was that there were some customers paying us even though the accounts were closed."

"I don't get it. What's the problem?" asked Marta.

"The accounts were closed for nonpayment. But those customer accounts were current."

"Okay, so we owe them a refund. I still don't see the problem."

"Most problems were minor. When customers complained, we apologized and gave them a free month or two. They were happy and life went on. But here's the scary part. I don't know how to say this."

"'Start at the beginning, continue to the middle, and stop at the end,'" said Jim.

"Alice in Wonderland," Denise smiled.

Jim started to speak again but Marta stopped him. "Tell us the rest, dear," she said.

"Some customers didn't complain. And the reason those customers didn't complain—" Denise hesitated.

"Go ahead, Denise," Marta prompted gently.

"—is…they're dead. They died. Their meds were cut off and they died. And I think it was done deliberately."

"You're kidding," said Marta.

"No." Denise picked up her glass and sipped her wine. She looked around. The shadows outside had grown longer as the day ran out. People hurried by on the street. They were like streaks of color flashing across the restaurant's window. Denise studied her wine glass as if there were an answer there to the riddle she'd found.

She shook her head slightly and refocused on her story. "I dug a bit and looked into the patient backgrounds to see if there was something they had in common. Maybe that would identify an error in the accounting system. And I found it."

She picked up her glass again and drained it. "Not one of them had any family to speak of. No husbands, no wives, no kids or parents. I couldn't even find any friends. Nobody to miss them. Dr. Cruz, Marta, I'd swear that these customers were selected because nobody would ask questions. It's just too much of a coincidence."

"Holy crap," said Marta, who never swore. "How long?" she asked in a clipped voice.

"The first case I found was a SNAP user named Emery Miller in Venice, California, about a year ago. Since then, I've found eleven other customers who had their nanoagents terminated for non-payment. Each one was from a different division of NMech. None of the deaths looked suspicious, so there was no investigation. But we're still getting paid. So the problem is not with the accounting programs, but with someone tinkering with the program, someone who's smart, but not an accountant."

They stopped eating while to absorb the news. Jim waved off a waiter who hurried to the table to ask if there were a problem. Marta picked up a wine bottle. "I think I need another glass. Anybody else?" There were nods around the table and Marta poured.

"That was about a year ago, you say?" asked Jim.

Denise nodded.

Marta and Jim looked at each other. Marta said one word, "Eva." Jim nodded slowly and said, "That would have been about when Eva was getting the bid ready for Rockford. Do you think that there's a connection?"

Movement stopped around the table. Denise looked puzzled,

but realized that Marta and Jim, even Dana, knew something that she was about to learn.

The waiter served the main course family-style. Beef stew served in a heavy kettle, accompanied by a delicate chayote squash and fried plantain slices. They pondered Denise's revelation while they ate. Dana only pushed his food around his plate.

Marta turned to Denise. "Can you make a list of the customers who were affected? We have to deal with this."

Denise looked miserable. "No. I can't. I was locked out of the system two days ago. I thought I'd been fired but I'm still on the payroll. Just all of my company access is gone."

"What the hell is Eva up to?" Jim asked. There was no reply.

❀　❀　❀

The NMech jet circled Boston's Logan airport until the air traffic controller indicated a break in the commercial traffic and provided landing instructions. The pilot taxied to a private hanger and rolled to a stop. Rafael Cruz and his escorts were met by two more NMech security agents. He was frisked and warned again.

A woman's voice said, "You're coming with me."

Rafael turned and saw a small woman. She directed the security men to flank Rafael Cruz, and then waved her sleeve at the ex-prisoner.

"Recording. Say hello to your daughter. She'll get the datafeed soon."

Eva Rozen's Boston home resembled her office—functional and unadorned. The dwelling's front door led to a stairway. At the third floor there was a narrow hallway that ran the length of the unit's spine. The lighting was dim and consisted of old-fashioned light bulbs. There were no brightwalls here. She'd even removed all of the windows in the apartment and replaced the self-cleaning, insulated nanocoated glass with old-fashioned window panes. It had been difficult to find a glazier with ordinary panes, but Rozen had the resources to pay for the out-of-style glass.

The apartment had the same configuration as her childhood home. The first room off the narrow hallway was a small bedroom, unused. This would have been Gergana's room. Next was the bathroom—cramped by the standards of Eva's current wealth, but one that matched the dimensions of her childhood apartment. Then a small bedroom, just large enough for a standard-sized box spring and mattress with ordinary sheets, a thin blanket, and a pillow. Next came the master suite and, finally, the kitchen. That was reduced to a small cupboard and refrigerator, stocked with an assortment of the humble foods from her childhood: blood sausage, spicy salami, vinegar-dressed potato salad and mish mash—an olio of vegetables, eggs, cheese, and spices.

The master suite housed the sole concession to luxury, a smartbed. It was king-sized, ironic given Eva's stature, and appointed with nanofiber sheets that were as frictionless as graphite and touched her skin as lightly as a whisper. The smartbed adjusted to her fidgety slumber and matched her body temperature, degree for degree. Despite the luxury, she slept no more than three or four hours at a time.

The black-clad NMech security agents who escorted Rafael to Eva's apartment spent little time observing their CEO's odd decorating sense. She had used them often as bodyguards, and, on occasion, for special services of a more intimate nature. They delivered Cruz to the guest room. One of the agents subvocalized a quick command to the apartment datapillar and explained to Rafael that he was to remain in the guestroom. He was not to wander anywhere else in the apartment, save the bathroom, nor was he to attempt to remove the security collar unless he enjoyed considerable pain.

"How long am I going to be here?" he asked.

"Don't know."

"What about my daughter? Can I see her?"

"Don't know. Stay put." They guards rechecked Cruz's security collar and then left.

Rafael sat down on his bed. It was even more uncomfortable than it appeared. He paced along the room's length and looked at the bare walls. He'd had more freedom in prison.

❀   ❀   ❀

The waiter brought coffee—Puerto Rican coffee, of course. "Our coffee was once considered the best in the world," said Marta, proudly.

"Right, Mom. Everything is better in PR. Is this from Yocahu, too?"

Marta smiled at her son. "Dana," she said with a gentle intensity. "Every growing thing is a gift from Yocahu."

Dana had been watching Denise and looked thoughtful. "Mom, we need to get Denise out of Boston, away from Eva."

"Why?"

"Mom, don't you see? If Denise knows about whatever Eva is doing, and Eva knows that Denise knows, Eva isn't going to let Denise alone."

"So, she'll fire Denise. We'll rehire her."

"It's not that simple," said Dana. "Do you think that Eva will let the only person with some proof of what she's doing just walk away? Eva will, uh, get Denise out of the way."

Jim spoke, addressing Denise, "My son can be melodramatic."

"Dad! Listen to me! Ever since Eva took on the Rockford bid

there's been something wrong with her. I could see it even though you tried to keep me away from her. And every time I tried to talk to you about it, you would change the subject. You and Mom wouldn't admit it. You think Aunt Colleen really had a heart attack? The last thing Eva said to me was that she had some 'big mischief.' What if Aunt Colleen was just the beginning?"

Turning to Denise, Dana said, "There's something wrong with Eva. She's going to see you as a threat, and she's not going to let you just walk around knowing about what she's done. You've got to go somewhere safe."

"I can go home," Denise said. "I live in Melrose."

Marta nodded. "Dana, you're right. And Eva will find Denise in Melrose." To Denise, "My dear, I'm sorry, but you've stepped on a hornet's nest. She must know that you figured it out." Marta thought for a moment and then smiled.

"Denise, have you ever been to a rainforest?"

"You mean, like the Amazon?"

"Like that," said Marta. "There are rainforests all around the world, but the gentlest one is called El Yunque. It's the most beautiful place on earth, and I have family there you can stay with. No one will find you there."

She touched her datasleeve and called up a display and was about to make travel arrangements. Dana put his hand on her sleeve.

"Mom, stop," he said.

"Why? Abuela's family can take care of Denise."

"Mom, think. How's Denise going to get there?"

"She'll fly. I'll pay for the ticket." She turned to Denise. "Don't you worry—consider this a work assignment. NMech will pay for your travel, and your time."

"That's just it, Mom. Eva's going to find out. You're still missing the point. Eva may be the richest woman in the world, but right now she may be the most dangerous person in the world. Let me do it. I can jack the airlines and get her on under another name."

"Since when does my son jack anything?"

"Mom, I'm almost sixteen. I know as much about ghosting as Eva does," he boasted. "Remember—she used to teach me. We kind of covered a little more than most kids."

"How long have you been ghosting?"

"Can we talk about it later? Right now, let's get Denise to Puerto Rico."

"Puerto Rico?" Denise exclaimed. "I'm going to the Caribbean? You mean it?"

Dana turned to her. "Make up a name. First, middle and last."

"Okay." Denise thought for a moment. "How about Simone Ann Bening?"

"Where did you get the name?" asked Dana.

"After the Flemish artist, Simon Bening. I just borrowed it."

"Better avoid a name from history. Eva will be looking for you already and her pillar will do a wide search. The searchbot will notice any coincidence and follow up on it. Let's make it, uh, Barbara. Barbara Anne Benning. Anne with an 'e'."

"OK," said Denise. "Barbara Anne Benning, Anne with an 'e' it is."

Dana held up his hand. "Link your sleeve to mine." She mimicked the gesture. Dana called up a display and subvocalized for a few moments. There was a half-second electronic conversation between the two sleeves.

"Okay," he said at last. "You're travelling as Barbara Anne Benning. Take the maglev from the South Station depot. You'll be in Philadelphia in about two hours but you have to leave now. There's a John Jays one block from the station there. I doubt that Eva will look for you in a high-end store that far from Boston. Buy yourself a carry-on bag and some summer clothing. I'll use my ghost to link to Mom's family in Puerto Rico and let them know they have a special guest on the way. They'll pass the message to Abuela, Mom's grandmother. You're going to love her."

"Dana, I can't afford John Jays," Denise said.

"Don't worry. You have an open account there now. Don't try to link with Mom or Dad because Eva will find you."

"Can I call my neighbor to take care of my cat?"

"Yes, but don't say where you're going. We'll deal with Rex later, after this is resolved. Until then, you can link with my ghost account. It's already on your sleeve. Anytime you link to me, start by saying, 'Abuela says hello.' Don't trust anything you think is from me unless I start by asking about Abuela's health. If you're in trouble, say that Uncle Roberto says hello."

Marta interrupted. "Dana, my uncle died three years ago."

"That's the point, Mom. You know that and I know that, but Eva won't because that part of your family doesn't use pillar-and-sleeve tech."

There was a hurried round of hugs. Barbara Anne Benning hailed a cab for the train station. She turned to the family that had befriended her. "Remember this. It's important. If you find the pillar that Eva is using to control the NMech accounts, look for some code that would put a hold on customer accounts for non-payment. Look in the accounts receivable programs. Normally, it's the credit department that places a hold. But look in receivables and you'll find her backdoor into the system. And thank you for everything."

She turned and looked at Dana. "If you were about ten years older…"

He blushed.

Then Barbara Anne Benning, née Denise Warren, stepped into a cab and disappeared into the Boston traffic.

Marta looked at her son. "I'm proud of you, but when this is over, we're going to have a little talk about ghosting. Let's get home now. I've got something that will help."

⚛ ⚛ ⚛

Eva arrived home three hours later. Rafael called out, "Hello? Somebody here? I'm hungry. Can I get out?"

Eva walked to the guest room-cum-cell. "Hold still," she said. "I get you something. Later, you will see your daughter. Maybe. Do what I say and Marta and your grandson will be okay. Don't cross me or all three of you have great pain."

Eva left and returned with food and water. "Eat up. I've got work to do."

She returned to her office and thought for a few minutes. How the hell did that accounting clerk stumble onto Cerberus? What did she tell Marta and Jim? This on top of the Rockford investigation? *I need that complication like I need a stump.*

Eva started to pace. Her arm itched again. She put on a piece of medical cloth to deaden the sensation and to repair the skin where it had been rubbed raw by her scratching.

"I need to hold them back for a while." She was talking out loud, addressing no one in particular. She touched her datasleeve. "This will do quite nicely."

A status light on the datapillar she called Cerberus turned green. She called up her display and subvocalized. Then the light turned from green to red.

The Great Washout had begun.

28

THE GREAT WASHOUT

BOSTON, MASSACHUSETTS

WAZA NATIONAL PARK, CAMEROON

PARAGUANÁ PENINSULA, VENEZUELA

BADULLA, SRI LANKA

MARCH 4, 2045

Halfway to their home in the Boston suburb of Brookline, Marta's sleeve pinged an incoming link from Eva Rozen. It was tagged "urgent."

"I just got a link from Eva," said Marta. She reached for her sleeve but Dana put a hand out to stop her.

"Wait until we get home," he said. "Whatever she wants, let her stew. She's had plenty of time to plan. Let's figure out how to respond."

Ten minutes later, they arrived at a rambling Federal-style home in their Pill Hill neighborhood, a two-story white house with black shutters. Fir trees dotted the front yard. The driveway passed the front door and dog-legged back to a large, well-maintained garden, now lifeless in the Boston winter. Theirs was one of the first homes built in what had been farmland nearly four-hundred years earlier. A wooded area abutted the residence, and beyond that, the ponds, brooks, and culverts that connect the Muddy River to the Charles River.

They left their scarves and coats in the mud porch and headed for the living room. Dana touched the wall and pressed gently. The walls, ceiling, windows, and floors radiated heat and the room was comfortable in moments.

Floor-to-ceiling windows covered the living room's length and offered spectacular views three seasons of the year. Today the winter view was dreary. A walking trail through the wooded area behind the house looked like a ragged streak of mud drawn across the frozen landscape. There were no robins, no crocuses, no tender green shoots. The first signs of spring were hiding, well aware that Boston winters could last for months. Snowstorms in April were not regular but not uncommon.

The family sat on chairs arranged in a grouping around a low, oval-shaped walnut coffee table.

"Mom, quarantine Eva's message before you open it," said Dana.

Marta pointed her sleeve to the pillar and transmitted Eva's message. The pillar would sequester any suspicious data to ensure the integrity of their sleeves and the house systems.

"My son, the security expert," Jim grinned.

"Dad, it's what I do. Let me open the file," said Dana. He stared into a heads-up display and began to subvocalize. "It's a vid feed. I don't see anything hidden in it but I'm going to have the dumb pillar display it just as a precaution."

The dumb pillar was not connected to any house systems, or to anyone's sleeve. Its function was entertainment, to project films, holos, vids, and music. Dana subvocalized again and the pillar emitted a beam of light. The rainbow holographic transmission focused in the center of the room. The image was a bit grainy suggesting that the recording was created on the fly. A plain room appeared, with a simple bed in the background and a man of moderate height in the foreground. His mahogany brown skin, black eyes, and salt and pepper hair looked out of place in wintery Boston. Dark wrinkles were evidence that he had spent years in the sun without anti-UV enhancements. He wore a simple cotton tunic, a security collar—and a frightened expression.

The man in the recording was looking ahead. "I remember you. You were with my daughter. Is she okay? Is that why I'm here?"

They heard Eva's voice, "She's fine. You see her soon enough." Then the field of view expanded. Jim and Marta and Dana could see two black clad NMech security men flanking the man in the video.

"Oh," said Marta, very quietly. "Is that my father?" She stared at the holo for several long moments and burst into tears. "*Dios mío!*" That's my father! How? I don't understand. He's supposed to be in prison." She started to crumple. The stress of the past several days had taken its toll on Marta's health.

Jim and Dana rushed over to catch her. Dana pointed with his head and said, "The sofa. Put Mom on the sofa."

The vid feed of the holo cut off. Eva's features replaced Rafael's. Her voice was strained, agitated, her speech reduced to simple thoughts. "Marta, you owe me. You owe me lot. I keep Jim out of jail. I make you rich. I help your poor. I get your father out of prison. Now he is here. You must do what I say. I mean it."

They watched in sickened horror. Eva had been friend, mentor, and colleague for years. She'd been a difficult friend, to be certain, but she maintained a unique brand of loyalty. Now she was changed. Dark circles ringed her eyes. Her hair was unkempt, unwashed. Her recent tics, jitters, and odd mannerisms had progressed to jerky movements, nearly uncontrolled, as if she were a marionette in the hands of a palsied puppeteer. She alternated between brushing non-existent bits of lint from her clothing and scratching hard on her left arm.

Eva's voice rose. Normally flat and uninflected, it was shrill and unsteady. "Forget police. I stop them anytime. You blame me for Rockford? Soon NMech gets Rockford. We get everything. I reorganize NMech. I get rid of waste. Stay away or I hurt your father. Stay out of my way."

The link ended abruptly.

"What the hell?" said Jim.

"My father," said Marta.

"He looked scared, but healthy," said Jim.

Marta's eyes welled with tears. "I've lived with the fact that I might not see him again, at least not for another decade. But that was him. If Eva got him out of prison now, why didn't she do it sooner? When she was still, well, sane?"

Jim wrapped his wife in his arms. She buried her face in his chest and sobbed.

Dana stood quietly. When his mother's cries subsided, he walked to the tall windows and stared out. Without turning back, he asked, "I wonder where Eva has him. Did either of you recognize where the vid was shot? Did anything look familiar?"

"No, nothing," said Marta. "But did you notice Eva's speech patterns? The syntax? Even her accent is returning. I don't like this one bit."

"There's something about the vid I can't quite put my finger on," said Dana. He turned away from the sad view of dirty snowdrifts and mud—and touched the window to darken it. He said, "Let's see the vid again." He subvocalized and Rafael Cruz appeared once more. They could see off-white walls in the background, the corner of a bed and the edge of a window. Once more, Eva delivered her tormented edict.

"She looks terrible," said Marta. "Her left arm is bleeding. She's scratching herself raw." Marta subvocalized and accessed information in her medical database.

"Why is she doing that?" asked Dana.

"Some of the medications for personality disorders can cause itching. I'm guessing that she's self-medicating in some way. Maybe it's induced some kind of mood disorder, like BPD."

"What's that?" asked Dana.

"Borderline personality disorder. It's a prolonged disturbance of the personality. A person with BPD can experience mood instability—"

"That's our Eva," said Jim, "But she goes a wee bit beyond instability."

"And it goes way beyond moods, Jim. Listen, we're dealing with Eva at her worst. We're in for a rough ride. Eva can't handle the emotions she's feeling. They're too complicated and too threatening. So she splits her feelings. It's easier to see things as all bad or all good. She idealizes herself, exaggerates her positive qualities, then devalues others. They become the 'all bad' to match her 'all good.'"

"Why now?" asked Jim.

"I can't even guess, but I can tell you that if we push Eva too hard, she could tip. She would demonize us. If she sees us as all bad, it will be easy to devalue us, to make us non-persons. Then she would have no compunction about killing us."

"Mom, do you think she killed Aunt Colleen?" asked Dana.

"I'm sorry, but yes, she might have."

"But if she killed Aunt Colleen..." said Dana, his voice trailing off.

"Remember, we're not dealing with a sane person anymore," said Marta.

"She was always nice to me," countered Dana.

Jim said, "She's never been normal."

"She was *always* nice to me. She was okay in her own way until you and Mom teamed up against her."

There was an uncomfortable silence. Jim spoke, "Dana, the Eva you knew when she was your teacher is not the Eva we're seeing today," Jim said.

"Why would she turn on us? Would she hate me now, too?"

"Dana," Marta said gently, "something happened to her, something changed her."

"Mom, I've been trying to tell you and Dad for a long time. But you kept on saying, 'Oh, that's just Eva,' or 'She can be moody.' But it all came down to public health. Without her all that would have been impossible." Dana choked back a sob. "And then you made sure that I couldn't spend any time with her. One of my best friends—and I needed a chaperone to be around Eva. Maybe I could have helped her. *You* could have helped her. You have all these herbs and plants from *Yocahu*"—he spat the word—"and you could have helped her."

Marta started to cry. Dana's accusation rang true. "Hijo mia, come here."

Dana held his mother at arms' length. He held her tenderly and respectfully, but at a distance. There was a mixture of pleading and command in his voice, "Mom, you have to figure out what happened to her so you can fix her. I like plain old weird Eva." Then he embraced his mother and they absorbed strength from each other.

But Dana wasn't finished. He turned to his father. "And you? She was your best friend. Then you didn't want anything to do with her. I don't know what she did to you, but can't you forgive her? Look where we are now. Why did you shut her out, anyway?"

"It's a long story, Dana," said Jim.

"Baloney! 'Long story' is what you say when you really mean, 'I screwed up.' Okay, you don't want to tell me why you stopped being her friend after all these years? Fine. But you took her away from me. And you had no right to mess with *my* friends.

The three stood silently. As one, they reached for each other, huddling together, each mourning in his or her own way. Jim shed silent tears. Marta sobbed gently. Dana stared straight ahead, angry one moment, crushed the next.

At long last, Jim spoke, his voice tactical. "Right now, we have three related problems. First, where is Rafael? Second, how do we get him away from Eva? Finally, there's the small matter of what she's going to do next."

"Play the vid again, Dana," Marta instructed.

They watched it again and then Marta pondered out loud, "This is troubling. My dad is one thing, but I'm terrified about what she's going to do. She said, *'I reorganize NMech. I get rid of the waste.'* I wonder if she's going to do something with our public health projects. That was how she described them—a waste. She said over and over that she only did it to get me to join NMech. My god, if she terminates the public programs, there are people who will die. I mean thousands, maybe hundreds of thousands."

"What kinds of projects, Mom?"

"Ah, a lot. Let me make a list."

Marta touched her sleeve and subvocalized. She frowned, and tried again.

"Guys. We've got trouble. I can't access any of the public projects, the subsidized patients, the donated nanomeds—none of it. I can't tell if she's terminated those programs or just locked me out. She could wash out every charitable activity we've built."

"How do we stop her?" asked Jim.

"I don't know. I don't even know what she's doing. I don't know how she terminated the accounts that Denise Warren told us about. I don't know how to stop her."

"Can the public health projects be restarted?" asked Jim.

"I don't know that either. I don't even know if NMech still exists. Oh God, I feel helpless."

"Dana, link to the newsfeeds. See if there's anything," said Jim.

Dana subvocalized and a series of images projected before them.

"Look. There. And there," Dana pointed.

One feed showed a panorama of hospital entrances, flooded with ambulances. There were desperate fathers and screaming mothers carrying their children. Old men and women with labored breathing, their faces pale or jaundiced. Another series of feeds showed the chaos and violence of street riots or worse—running battles between military or police agencies on one side and pirate armies on another.

Dana stood, speechless. Marta sat down heavily, her legs unable to support her. Jim watched the feeds. Dana moved to comfort his mother.

"How is she doing this?" Marta cried.

Dana asked, "Can she be controlling this from her office?"

"I don't think so. Eva would have hidden everything. I had the technical staff search for anything that looked suspicious right after the explosion. Eva's pillar had been dormant. She must have another one somewhere."

"I bet that's where she's got my grandfather," said Dana. "I think I saw a clue in the vid. I'm going to play it again to be sure."

❀   ❀   ❀

Morning broke on March 4. The usual chitters, howls, and grunts of Waza National Park's wildlife were joined by a new sound. Cries of dismay and alarm echoed among Sergeant Mike Imfeld's squad. Their uniforms were dead. It was as if a master switch turned every uniform to dumb cloth. The medical sensors were muted; the protective liquid armor puddled uselessly; shirtsleeve bandages for cuts or scrapes morphed from medical marvels to blood-mottled fabric. Even the command, control, and communications applications were dead. In the event of an attack, they would be reduced to blind firing.

Imfeld's problems were compounded by his foe's skill. Aluwa's scouts had come of age in the forest. A small company followed Imfeld's squad's every move. Seventy-five child-soldiers circled north above Waza and then south to reach the park's eastern border and set up an enfilade with a company on the western border with Imfeld's squad in the middle. Aluwa knew he would have the element of surprise. What the teenaged general did not know was that the EcoForce's defenses had been disabled by instructions from Cerberus.

When Aluwa's attack began at 0700 hours, local time, the EcoForce squad's defense was unfocused. The battle for Waza National Park was over in less than thirty minutes. Aluwa suffered nine casualties. None of Imfeld's troops survived. The Great Washout claimed its first military casualties.

❀   ❀   ❀

Dana Ecco subvocalized and the dumb pillar projected Eva Rozen's vid feed. There was Rafael Cruz. Behind him were the plain walls, the small bed and the edge of a window.

"There," said Dana. "The window."

Jim said, "So what? We can't see what's outside of the window."

Dana said, "Look at the window itself. That's not smart glass. It's an old-fashioned window. Look at the glass. See the little ripples in the window?" He subvocalized and magnified the image which showed a moiré pattern in the glass.

"You're right," said Marta. She stared at the vid. "Nobody uses plain glass anymore. Building codes require smart glass."

Dana said, "So where does Eva go that would have this kind of a window?"

"I bet it's her home," said Marta. "I remember, back at Harvard. One of the few times she ever talked about her childhood, she described the apartment where she grew up. She swore that if she ever was successful—no, make that *when* she became successful, she never had any doubts—she wanted to recreate her childhood apartment in Sofia."

"So what do we do now?" asked Dana.

"We're going to pay her a visit," said Marta.

"Not yet," said Jim. "First you and Dana go to her office and see if you can find anything that will restart the public health programs. I'm going to get your father."

Marta said, "I think she's taken on some enhancements. If I'm right you can't face her without being prepared. I'm expecting a delivery, something for you that Eva won't expect. And I need a little time in the lab to confirm my suspicions about her."

⚘ ⚘ ⚘

When the Cerro Rojo plant failed, Nancy Kiley made an inventory of the region's available water, took stock of her own situation, and made an executive decision: she fled.

The region's principal water reserve was the eight million gallons remaining in the pipeline that carried Cerro Rojo's output to its customers. Kiley did a quick calculation to estimate how much time she had. Eight million gallons of water for thirty million thirsty people. Fifteen quarts each. Survival ration for a healthy human at rest is a bit over three quarts daily. If they used the water in the pipelines only for drinking, the region's population could survive for a few days—if it rested in the shade. Factor in sanitation and hygiene, the need rises to fifty quarts a day or more. But agriculture and industry also lay claim to the liquid treasure.

She had little time. Within hours, the populations of six Caribbean islands and the northeast coast of South America would be parched. And Nancy Kiley wanted to be anywhere in the world other than in the middle of a water riot.

Water, water, everywhere, Nor any drop to drink. Nancy Kiley stuffed her travel documents, a change of clothing and a few personal items in to a bag. Before she commandeered an NMech land vehicle, she doubled back to her tent and stuffed a treasured pair of comfortable shoes into her pack—*I'll be damned if I have to wear freaking boots when I'm out of this shithole.*

"What are you doing, Dr. Kiley?" her administrator asked as he watched Kiley leave. *I can be in Maracaibo in less than two hours, but not if I have to take time to explain.* "It looks like there might be a problem upstream of the plant," Kiley temporized. "I should be back in a few hours."

"You can't leave now. What are we going to do here? The whole system is down. What do we try now?"

"You've got a whole team to figure it out. Stop complaining and get to work." Startled, the admin turned and left.

Kiley went back to her escape plan. Canada had been spared the worst of the drought and its climate harbored none of the fire ants, scorpions, and the other creatures that had bedeviled her here in Paraguaná. She could walk on pavement, not gravel, and enjoy seasons with temperatures less than 90 degrees. If she could make it to Toronto, then she might be safe.

Kiley subvocalized and checked airline schedules. Bad news. No flights to the northern United States or Canada until the next day. By then riots would overtake the airport. She started to weep. *Goddam Eva Rozen for getting me into this. Why don't you come down here for a spell, you fucking dwarf!* She pounded the dashboard in frustration, then chided herself. *Come on, Nancy—think like a scientist, an executive.*

An executive? That was an idea. She wasn't part of the most senior management, but perhaps she could appropriate one of their privileges. *Eva spared no expense to get me here. NMech can spare no expense to get me the hell out again.* She found a corporate jet in Boston, fueled and idle after a flight from Mexico. Kiley linked to the pilot. He was agreeable. There were no travel orders from Rozen and anyway, she'd been unreachable. Nancy agreed to a fare equal to a month's salary and the promise of some personal time with the pilot. He would be in Bogotá when she arrived and would take her to Canada.

Kiley left her vehicle at the Maracaibo depot and sprinted to the maglev. The region would soon be bloody, but with a little luck she'd be airborne before it all went bad. For the first time in weeks, she began to relax. A long shower topped her list of things she'd do when she was safe. No, make that a bath. Hot water up to her chin. Quiet music, a bottle of wine. Make that two bottles. She would soak till her skin was as wrinkled as a prune.

The maglev decelerated at the airport and Kiley came out of her reverie. She grabbed the pack with her travel papers and clothing and headed for a private terminal where the NMech jet was fueled and ready. Fifteen minutes later she was pressed back in her seat as the aircraft accelerated. The landing gear bumped as it folded into the belly of the craft, and after a steep banking turn into the sun, they were heading north. Within minutes she and the pilot were cruising at 30,000 feet, destination: Toronto. There she would find cool weather, moist air, no water shortages, and no damn bugs.

Nancy Kiley unbuckled her seat belt and stretched. From the plane's bar she poured vodka into a tumbler of ice and swallowed half, cherishing the cool burn in her throat almost as much as the quiet roar of the jet. She shut down her commlink. Let her staff, no, make that her former staff, let them deal with the desal plant. For the next few hours, Nancy Kiley would enjoy the solitude of the plane's small but comfortable cabin and its well-stocked bar.

She freshened her drink and took her pack to the lavatory. There was a shower, large enough to lather and rinse. She drained the glass, stripped, and stepped in. The water was tepid. *As long as I'm away from Cerro Rojo, I'd shower in a glacier,* she thought.

Nancy lingered, lathered, rinsed, stepped out of the shower and toweled dry. She poured another vodka, her third. Her clothing was stained despite the self-cleaning nanofibers. No matter. She would buy a new wardrobe in Toronto. She shrugged into clean bra and panties from her pack, along with a fresh tee-shirt and slacks. Nothing fancy, but clean.

Her one nod to fashion was the shoes, a pair of Dolce & Gabbana ballet flats, shoes that she'd carried halfway around the world. Nancy handled them with the reverence reserved for a holy relic. They were comfortable, lace print silk and leather with a tiny version of the distinctive D&G logo worked into the print pattern. Kiley smiled in anticipation. There hadn't been an opportunity to wear D&G in the Paraguanán scrubland.

The shoes. Something about the shoes. What was it? Fatigue, dehydration, altitude, and alcohol slowed her thinking. She giggled and reached again into her pack. Where were her socks? *Well, Dolce*

& Gabanna was made for bare feet. She steadied herself, sighed, and slipped the left shoe onto a tired foot.

Had Nancy Kiley been sober, she might have looked inside the shoe before slipping it on, out of habit, or to admire the fine Italian workmanship. She would have seen the bright yellow amphibian, smaller than the tip of her thumb, enjoying the cool darkness in the toe of her shoe. A sober Nancy Kiley might have found her socks. The material's tough nanofibers would have repelled the frog's poison. Even after direct contact, Kiley might have survived were it not for the cracked skin on the bottom of her feet.

The stowaway was a female Golden Dart Frog, *Phyllobates terribilis,* reputed to be the most poisonous of the area's small amphibians. Its skin accumulates a cardiotoxin that leads to convulsions, swift and certain. Brilliant markings warn predators—a caution that Kiley would have seen nine ounces of vodka earlier.

At first, Kiley felt a warm, rubbery sensation. Then pain. The Golden Dart's toxins offer an unpleasant death, mitigated by hallucinations and by the speed of the poison. A few minutes of agony and disorientation for Nancy Kiley, then oblivion.

When the flight landed in Toronto, the pilot taxied to a private terminal. Once the craft's engines were silent, he went into the passenger compartment, and halted abruptly. He stared, uncomprehending. His passenger, quite dead, was curled in a fetal position, wearing a shoe on her left foot and clutching her right shoe in a rigid fist. The pilot recoiled in panic, then giggled uncontrollably and recited an old nursery rhyme. "Deedle, deedle, dumplin', my son John. One shoe off and one shoe on." He shook his head to clear his thoughts. He had just landed an unauthorized flight carrying a dead body into a foreign country. He subvocalized to ready his flight back to Boston and noted that martial law had been imposed in the larger Caribbean islands and that over a thousand civilian casualties had been recorded in the first few hours of the Great Carib Water Riots.

❀   ❀   ❀

Jagen Cater stumbled out of the train's lavatory and took hold of the top of each seat he passed to steady himself until he fell back into his own seat. He closed his eyes in resignation. The face he'd seen in the mirror was jaundiced. The task of urinating had become difficult and what he saw had terrified him. His urine was cloudy with waste. Its frothy presence in the toilet told him that his dialysis device had failed.

Now he understood the exhaustion, the disorientation. All of the symptoms of end-stage renal failure were present: fatigue, confusion, swelling of the feet and hands. No wonder his shoes felt too tight. Even bad breath—hadn't the conductor shied away from him? Next would come the nosebleeds, the bruising, the bloody stools and urine. His hands and feet would become numb. Walking would be difficult. Confusion would peak just before he lapsed into a coma.

With stoic fatalism he reasoned that the IDD had given him five years of life. If it were his karma to leave the material plane today, then so be it. *Too bad about the fine plucking. The year's harvest would have been superb...*

Before this day was over, Jagen Cater's thoughts would turn to the Compassionate Buddha. His invocations would be joined by the prayers of some half million other IDD users—invocations to Jesus, to Allah, to Krishna, to the Great Spirit, to a higher power. All would fall on unresponsive ears.

Eva Rozen's Cerberus program was implacable and denied

their appeals for life, turning the murmured pleas into prayers before dying.

❁   ❁   ❁

Eight thousand, five hundred, seventy-seven miles east of Jagen Cater, one third of the world's circumference, a disheveled Eva Rozen paced. She was jittery, her movements awkward. Her hair, normally a tight mass, was tousled. She picked repeatedly at her rumpled clothing, pinching nonexistent bits of lint. For the past two hours, Eva's assistants, public relations in particular, had tried without success to link to her. Proposals to review, contracts to approve and plans to implement, were ignored. Eva's thoughts were far from NMech operations.

Her focus was internal. Images below the level of her conscious awareness pressed insistently. Unconscious memories vied with immediate needs. The push and pull of the past, set against the demands of the present, was taking a massive toll on her equilibrium. She couldn't concentrate. Her arms continued to itch. She was speaking to herself—or was it to an unseen audience? The patter was unintelligible.

For the first years of Eva Rozen's life, Mama and Papa treated her first as an object of pride and then as one that inspired disdain. The child Eva looked into this mirror and found herself both worthy and contemptible. Her self-loathing grew despite the nurturing she received from her sister. Eva developed survival strategies—aggression and an exaggerated sense of entitlement to bolster a fragile ego. Strike first, lest ye be stricken; strike harder if ye are struck.

For four decades, she balanced antagonism with an unrelenting need for acceptance. She found this acceptance in only three people. The first was her sister, Gergana, who loved and sheltered and comforted the infant Eva and the juvenile Eva. The second person who had accepted her without reservation was Jim Ecco, whose marriage to another Eva facilitated. And she found acceptance from Dana. Eva idealized him as the child she could have been, the life she could have lived, the child she could have borne, until Marta ripped Dana out of her orbit.

The weight of memory, the inexorable pull of longing unfulfilled became unbearable, and the structures that supported her rational mind collapsed. The mechanisms that filtered the bewildering din from the Table of Clamorous Voices were swept away, and with them Gergana's murmurs of comfort and adoration were lost. The loudest voices—those of Bare Chest, Doran, Papa—held sway.

Eva Rozen began to decompensate and an unquenchable rage emerged.

❁   ❁   ❁

Jim and Dana waited while Marta excused herself to her home office and lab.

"What's Mom doing?" Dana asked.

Jim shrugged. "Either something with science or something with Yocahu. I can't always tell the two apart."

"Do you believe in Yocahu and Juricán?" Dana asked. He sounded nervous, as if trying to replace fear with philosophical speculation.

"I don't know. It works for her. And the herbs and plants that she's found also work. If your mom wants to credit Yocahu, then that's fine with me."

Their desultory conversation trailed into silence. Dana told his father that he worried about him. What would happen when he found Eva? "I can help you, if you let me, but you just think of me as a kid. Remember, Eva herself taught me computer science and nanotech."

Jim said, "We'll see," the universal parental response which means, "The answer is no but I don't want to discuss it now."

"What about Mom?" Dana pressed.

"She needs your help, Dana. The last few days have been tough on her." Marta's gait, already slowed by her long battle with juvenile rheumatoid arthritis, had deteriorated. Her range of motion narrowed and she hobbled more than walked. Her skin was flushed with a rash. This meant that her periodic fevers were raging once again.

A few minutes later Marta reemerged. "Let me tell you what I found. I started to wonder why Eva had what appears to be a psychotic break."

Dana interrupted, "What's that?"

"Loss of contact with reality," Marta answered. "A major personality change. Did you notice how her language had become reduced to simple sentences? Her language pattern suggests thought disorder. It's a clue to her thinking."

"She did talk strangely," Jim agreed.

"She also is extremely jumpy and excitable. Lately, she's even moving more quickly, almost like she had taken a stimulant. For a while I thought maybe she was using cocaine. It fit, except for the language patterns."

"Couldn't she be just plain nuts?" asked Jim.

"Yes, but how is she nuts? That's the question. And can she get back to normal?"

"Who cares?" asked Jim. "I mean, why not let the police take over? They're a helluva lot better equipped to handle her, don't you think?"

"She holds the key to restoring the public health programs, to undoing the damage. If she stays in a psychotic state, she may not be willing to help. She may not be able to help."

Dana asked, "Why'd she break now?"

"That's the key question and I think I found the answer. I grabbed her coffee mug from her office to test for stimulants, amphetamines, cocaine, and even SNAP but there were no traces of anything like that. On impulse, I tested the cup for traces of neurotransmitters and found incredible levels of acetylcholine, or ACh. That explains a lot.

"Neurotransmitters relay instructions from the brain to muscles," Marta said, warding off Dana's next question. "ACh governs the speed of your thinking and reactions. Not enough ACh and your brain slows. Remember the old syndrome, Alzheimer's disease? When a person loses her memory and gets confused? That was linked to low levels of ACh. My theory is that she attempted to use ACh to speed up her thinking and reflexes."

"Why would she do that?" asked Dana.

"I think she experimented on herself when Rockford ran so far behind schedule. Again, I'm just guessing, but I believe that she was hoping that an ACh boost would help her think faster and work faster. And we did get the bid in just on time."

"Fat lot of good that did," Jim said. "What went wrong?"

"Too much ACh, then you can become excitable and paranoid. I don't think that Eva knew how delicate brain chemistry is. Or she didn't care and was willing to bet everything on this bid. Her body must be under as much strain as her mind."

Dana asked, "So, does this, this neurotransmitter make her more dangerous?"

"That'd be my guess." She turned to her husband. "Jim, if I'm right, when you confront her, you have to be careful. She'll be paranoid and irrational. And you can count on her moving a lot faster than anyone you've ever encountered."

The conversation was interrupted when a messenger delivered the package Marta was expecting. Inside the bulky container were two formfitting pieces of nanotextile clothing.

"Okay," said Jim. "So, what's up with the skinsuit?"

"Dana and I are going back to NMech to search her work areas again to try to restart the public health programs. The skinsuit is for you. It may keep you alive."

❁　❁　❁

The Great Washout was about to award Eva Rozen with the credentials of a mass murderer, and a unique one at that. It was carried out singlehanded, initiated in seconds, and fulfilled in less time than it would have taken to read a roster of the dead. There were some 20,000 fatalities from the Caribbean water riots and another 120,000 dead from dehydration. Nearly a half million IDD users; 110,000 dead from diabetic convulsions. Lt. Colonel Fierra's dozen troops. Hundreds of recreational drug users. And millions of others were imperiled, including a close-knit Boston family that shared a passion for science and humanity.

29

THE FOURTH FLOOR

BOSTON, MASSACHUSETTS

TUESDAY, MARCH 4, 2045

Commonwealth Avenue begins at the western edge of Boston's Public Garden, under the watchful eyes of no less a luminary than George Washington. At thirty-eight feet, the mounted leader on his bronze steed is the city's largest public sculpture and perhaps the most impressive. The horse's eyes looked ahead. Its ears were pricked forward, drawn to a looming battle.

The boulevard leaves the tranquility of the Public Garden and stretches through Back Bay, Allston, Brighton, and Newton. It crosses the Charles River and continues west. Eva Rozen's home was a stately brownstone in an elegant and exclusive neighborhood just three blocks from George Washington's prescient horse.

A grassy mall dotted with statuary and memorials divides the avenue. A few months earlier, the mall had been sanctified in the first snow of winter. Today it was cold and raw, gray and forbidding. Jim's senses were on high alert as he approached Eva's home on foot, scanning the building with caution. He'd often strolled here, but in all the years Eva called Commonwealth Avenue home, she'd never entertained a single guest. Jim had long admired the rowhouse's sandstone facing. An eagle guarded the entrance, its copper green wings unfurled. *Today*, he thought, *it should have been a vulture.*

He walked west, away from the Public Garden, past Eva's home. He wanted a good look at the building without being noticed. He adopted the posture and gait of other pedestrians in order to blend into the street scene. He did not even turn his head, but examined the front entrance with his peripheral vision. The door's lock appeared to be keyed with biometrics. This would not be a problem but surveillance devices might be.

Jim continued west at a stroller's pace past Eva's building and turned right on Dartmouth Street. He crossed the mall that bisected Commonwealth Avenue and turned right again. There was an alley between Commonwealth Avenue and Marlborough Street and Jim ducked into its narrow passage. He shed his outer garments, touched his datasleeve and activated the skinsuit Marta had given him, pulling its hood up over his head.

She'd obtained it from a small company founded by a former NMech scientist who remembered Marta's kindness and was willing to provide the suit. "It provides invisibility and partial armor," Marta had explained. "The armor isn't as good as NMech's, but Eva can disable an NMech skinsuit. And it has a chest pocket so you can carry a few small items without compromising your stealth mode."

"How is it different from NMech's military garb?"

"NMech's smart fabrics use layers of light-sensitive plastic threads that copy the appearance of the environment. It's better than camouflage but it's not true invisibility. This suit uses a different technology that will render you invisible."

Jim took the unimpressive-looking suit. It was covered with a pattern of tiny hexagons that resembled a quilted mattress pad for a doll house. "You sure?"

"I hope so. It bends light using tiny crystals, stacked like a woodpile. Anything underneath the crystals is undetectable at both the visible and infrared spectrums."

"How does it work? Give me the simple version, please."

"It's an old technology, developed in the early 2000s. Back then, scientists weren't able to cloak objects larger than a tiny fraction of an inch. My colleague found that by building the crystals from nano-sized carbon molecules the cloaking effect would work on larger objects."

Jim played with the suit. "It feels like it would be comfortable. You say it's armored?"

"Partially. That's the tradeoff. Complete invisibility but the armor isn't as good as magnetic shearing fluid. It uses silicon woven into tiny hexagonal cells. Each cell transmits the energy of an impact to its six neighboring cells, and these in turn to twelve more cells. Then to the next eighteen, and so on. It spreads impact over the whole suit."

"I've never seen anything like this. Is it new?"

"No. The armor was developed years ago for sports gear and luggage."

"Wait. I'm facing Eva Rozen at her worst wearing a *suitcase*? Why don't I just use NMech military armor?"

"Eva can disable anything built by NMech."

"How come nobody's used this technology for body armor before now?"

"There's a downside. No matter where you get hit, you're going to feel the impact all over your body. Remember, each cell transmits the impact to each neighboring cell. The force will be reduced, but you'll feel it everywhere. I'm hoping that this works well enough to keep you alive."

That was two hours ago. Now Jim, invisible, approached Eva's home on Commonwealth near Clarendon Street. He had to weave in and out of pedestrian traffic to avoid collisions. *That's a drawback to stealth*, he thought.

Pausing at her doorway, Jim took a small aerosol can and sprayed an arc of nanoelectronics suspended in paint around the door. This electronic doorway would block any signals or electronic traffic from the entryway's security. As he sprayed, he was reminded of the old Bible story of the ancient Hebrews, preparing to flee their Egyptian slave masters. The Hebrews painted a splash of the blood of a slaughtered lamb over their doorways to protect the household from the Angel of Death. Jim hoped that his sign on the doorway would disable Eva's security measures as well as the lamb's blood protected the Israelites.

Next, he placed a metal ring below the front door's hand sensor. The fist-sized circle contained powerful magnets at the four compass points. He activated the instrument and turned it in a counter-clockwise direction. The magnets pulled the deadbolt free of the strike plate. If the deadbolt were crafted from a non-magnetic material, the ring would generate an electric current to power the lock's motor.

Eva Rozen was a brilliant chemist but she was not a security expert. The door opened to Jim's device in seconds. Once inside, he paused in the entryway. He hoped that Eva's determination to recreate the apartment of her youth would mean little enough security that he could find his father-in-law, stop Eva, restart the public health programs, and stay alive. *All in a day's work.*

The hallway's dark paneling lent a claustrophobic feel, and the unfinished pinewood flooring looked shabby. Jim was surprised. Eva's wealth would have allowed any extravagance, but this part of her home was dark and cramped.

He felt his way up a stairway. There was no sense being stealthy. Most likely Eva already knew he was in. Still, he tested his weight on the outer edges of each step to minimize the groans of the old timbers. Up a second flight to the third-floor landing. He saw a narrow hallway and counted five doors down its length and saw a cramped kitchen at the end. *This must be what her apartment in Sofia was like. It's amazing how much squalor you can buy when you're rich.*

Jim paused at each door and listened for several moments and then placed a room reader on the door. The card-sized device displayed any movement within, even the slow rise and fall of someone's breathing chest. It displayed the size of the room and the position of any occupants. It could zoom in on an object or take in the entire space.

There was nothing in the first three rooms. Jim sensed a presence in the fourth room even before he used the reader. When he did, it displayed a figure on the left side of the room. Someone was inside, sitting still. Jim enlarged and then focused the shapes within the room and saw that there was a man seated on a bed. The display showed a window on the back wall in the same position as shown on the Eva's vid. Jim tried the doorknob. It was unlocked. He deactivated the skinsuit, took a deep breath and opened the door slowly, wincing as the old hinges complained. The man inside was

Marta's father, Rafael.

Jim slipped into the plain room and closed the door behind him. "Sir, are you okay?" he asked, his voice low.

Rafael was wearing a simple white tunic and gray gabardine slacks. He had cheap canvas slip-on shoes. Prison shoes. There was a black band around his neck. He started. "Who are you?"

"I'm Marta's husband."

"Jim?"

"Yes, I am. Pleased to see you again, sir. Do you know if Eva is here?"

"I heard her go up the stairs," Rafael said.

"Is that a security collar?" Jim asked, pointing to the black band around his neck.

"Yeah. She said this thing will hurt me bad if I go anywhere in the house except the bathroom. Worse, if I try to take it off."

Jim said, "Let me look, see if there's any way to remove it."

He examined the collar and then touched his datasleeve and linked to Dana. The commdisk on his jaw vibrated as he spoke with his son.

"Where are you and your mom?"

"Eva's office at NMech. We can't find anything here. What about you?"

Jim said, "I'm in Eva's house. No sign of her yet, but I found your grandfather. Tell your mom that he's okay." He heard an exclamation of relief as Dana relayed the news.

"Look," Jim continued. "There's a security collar around his neck. I want to get it off. You have any ideas about how to disable this thing?" He held up his sleeve and captured an image of the collar and its schematics. A databurst transmitted it to Dana.

The link was silent for a few minutes. Then Dana said, "No.

It's got a fail-safe. It'll generate a high-voltage electric shock before you could get it off his neck. Maybe fatal."

"I suppose Eva can get it off," said Jim. "I just have to convince her."

"Dad, it's got a fail-safe. I don't know if Eva can get it off. I think it's permanent."

"Shit."

"Yeah."

Jim thought a moment and asked Dana, "What about modifying it? Can we make it harmless?"

Dana was silent again, pouring over the information Jim's sleeve had transmitted. "We can try to lower the output, maybe make it non-lethal."

"Can that backfire?" Jim asked.

"What?"

"Backfire."

"What's that?"

"It's a term from when cars had spark plugs."

"What's a spark plug?"

"Never mind," said Jim. "Could it go off by accident when we try to lower the power?"

"I don't know. Maybe," said Dana. "But do you have any choice? Otherwise, he'll be stuck in that room for a long time."

Jim had broadcast the conversation so that Rafael could follow. The older man said, "I don't want this thing on me at all. I was better off in prison without it. Do whatever you have to do. I want to see my daughter and my grandson."

"Okay," Jim said. "Let's do it. I have a feeling that Eva's expecting me."

"Dad, I'm going to send you a file," Dana said. "Once you get it, activate the file then transmit it to the security collar. Let's hope that works."

"I'll keep my fingers crossed," Jim said.

"Why?"

"Never mind. Just send me the file." Jim's sleeve emitted a quiet chime. He had Dana's transmission. He subvocalized and then pointed his sleeve at the collar. Another chime announced that the file was accepted.

"Well, Rafael seems to be all right," Jim reported. "Sir," he said to his father-in-law, "I'm going to find Eva. Please stay here unless it's an emergency. I don't know what's going to happen with the collar."

He looked Rafael over one more time and then said to Dana, "This is it. I'm going to find her. When I, uh, resolve things, I'll link back. Now I'm going to link to your mom and then I have to go to work. See you soon."

Jim linked to his wife. "Querida, I'm going after Eva."

"Be careful." Her voice caught. "Te quiero." I love you.

Jim broke the connection.

He left Rafael in the small bedroom and reengaged his skinsuit. Approaching the stairway, he took a deep breath and climbed, flush with determination and dread. Sixteen steps to the fourth floor. He heard Eva pacing. He inched his way towards the sound.

This floor was different. The construction was new. The walls were paneled with a lighter wood, a reddish hue that gave a more expansive feel. Still, Jim felt hemmed in, despite the light and airy character of the timber.

The hallway led to a large, open work area. Sunlight streamed in through full-length windows. Unlike the windows in the rest of the house, these were modern nanoglass. The floors were ebony and teak. The woods were fashioned into a black and dark brown sunburst, centered in the middle of the room.

Jim heard her before he saw her. Her breathing was uneven and there was an odd crinkling sound as she moved, something like cellophane. It reminded him of wrapping paper on Christmas presents in his childhood. He remembered the barren feeling of the holidays, wondering what might anger his father.

He had not thought much about his parents for years. He'd sent a databurst link to them after he and Marta married. His father replied with an old-fashioned card, something that appeared to have been purchased from the stationary aisle at a grocery store. The card was white, with a silver pair of wedding bells embossed on the cover. The stilted poem inside the card appeared to have been composed by a journeyman writer. The prefabricated message started with the words, "We wish you years of happiness on your wedding day…" Jim wondered how he could enjoy years of happiness on a single day. The card was signed, "Your father and mother." Not, "Dad and Mom" or, "With love…" or even, "Best wishes…"

His father's scrawled signature was tiny, the writing faint. Jim could see places where he'd stopped and started. Marta said that the unsteady hand and uneven pressure suggested his health was failing.

"He couldn't have written, 'With love'? Or something personal?"

"Shhh…Querido. Let it be. His signature looks like that of someone with some neurological degeneration and loss of muscle control. Maybe Parkinson's. At least he sent you a card."

"Yeah, but he's still playing cock of the walk. He didn't even let Mom sign it." And then he never heard from them again, not once, until his mother was dying.

She died six years earlier. *Six? Seven? I don't remember.* An attendant at the hospital where she spent her last days had linked to him, to let him know he should come immediately. Her kidneys were failing and she'd refused dialysis. "She just wants to go," the attendant explained.

Jim flew to Pasadena in an NMech jet to make his peace with her. She was wan and drawn. She greeted him warmly at first, but within minutes, mother and son were arguing. It was as if no time had elapsed in the past quarter century. *I guess deathbed scenes work better in vids than in real life,* he thought.

The crinkling sound was louder and it pulled Jim's thoughts back to the present. He shook his head to clear his thoughts. Across the open studio Eva stood, feet close together. She seemed to sway. Her eyes were opened wider than normal and had a feral look. There was a rigid tension in her posture. He touched his datasleeve and allowed himself to be visible.

She spoke. "You come to me. You have to. You're more like me than Marta."

"No, Eva. This has nothing to do with Marta, or with you and me."

"You owe me. I helped you. I helped her. Stay here."

It was like hearing the petulant demands of a toddler. He tried to reason with her as he might with a child. "Eva, you're a great woman. You are good to your friends. I admire you. But what you did is hurting people, killing them. Tens of thousands of people, maybe more."

"Disregard that." Her voice was matter-of-fact, as if Jim had announced the weather.

"Eva, do you know what is going on around the world? The good things you created are falling apart. We can work together to rebuild it all. Eva, I am your friend and will always be your friend. Let's fix your good works before more people die."

He started to edge towards her.

"Don't come near me. I'll hurt you."

"I thought you wanted me to stay."

"Maybe you're not really my friend."

"Eva, please. Let me help you." Jim kept moving, an inch or two at a time. He averted his gaze, tucked his head down and hunched his shoulders slightly. Subtle transformations in body posture made him look smaller, non-threatening.

Eva took a step. There was that crinkling sound again. He looked carefully at his lifelong friend, now changed into...what? Her garments were covered with a network of black cables, each no wider than a blade of grass. They ran down her arms and legs and around her torso. Jim looked puzzled, then surprised. She was wearing an exoskeleton, electro-active polymer fibers that magnified her strength and allowed her to lift several-hundred-pound objects or strike with superhuman force. She began to walk forward. Jim held his arms out, palms up, as if to say, "I'm no threat."

Eva advanced. A look of rage had replaced her usual expressionless demeanor. There was no mistaking her intent.

❀   ❀   ❀

Marta and Dana searched Eva's office again. There were few papers. Dana helped his mother to jack the datapillar. It was another dead end.

"Do you suppose that she wiped the pillar of any traces?" Marta asked her son.

"There would be *something* there to find, I think. I'd bet she used a different pillar to wash out the public health programs. Maybe from a pillar at her home."

"Let's go help your father," Marta said.

"No," said Dana, "We'll be in his way. He's better off solo. And her pillar is probably protected. I want to look for something here that will help disable the pillar. So, let's wait till he links to us."

❀   ❀   ❀

Jim stood fast and spoke soothingly. "Eva, you have so much power. You can help. People are rioting in the Caribbean because there's no more water. Diabetics are going into shock. Kidney patients are dying. They're innocent. You have the power to save them. Then you and I can sort things out."

"You can't stop me," Eva said in a jittery voice and continued to move towards him. She picked up a chair. Jim started to react even before her hand touched the chair's straight back. He almost wasn't fast enough. She threw it at him with no more effort than swatting a fly. The chair hit the wall behind him and shattered.

"Eva. You don't have to do this. Let me help you."

She said nothing and closed in on him, a blur of motion. She lashed out with her right hand, a palm strike to the solar plexus with enough force to stop his heart. His skinsuit kept him alive although he felt the impact all over his body. Everything hurt. He flew backwards and crashed to the floor.

Eva came closer and then kicked him just above his knee. The force of the blow stunned him and he winced, but his leg stayed intact.

"Armor won't help," said Eva. She touched her datasleeve as Jim staggered to his feet. She aimed a punch. It was too fast to see, but he had started pivoting as soon as he saw her tense to strike. Not soon enough. Her fist hit his shoulder. He was thrown aside but unharmed. He might as well have been an empty container, tossed into the trash.

Jim started to his feet when he felt a wave of fatigue sweep him. He felt weary to his bones. *What's the use? I should have stayed out of her life. I should have stayed a dog trainer.*

His mind wandered again and he thought of Ringer. He missed her. The move to Boston had taken years from her life, and he had mourned every day since her death. Naively, he had thought that she would be happy anywhere she could have a walk and a toy. But the northern winters stung Ringer's eyes and matted her coat. It sliced her delicate foot pads with shards of ice and then burned the wounds when she walked on the rock-salted streets. Just going outside was uncomfortable. She'd pee on the back porch, where the snow had been removed, rather than walk down a short set of stairs and search for a snowless spot where she didn't risk frostbite on the business end of elimination. Eventually, he learned to shovel the apartment's tiny back yard for her, an exercise that provided a good deal of amusement to the neighbors.

Eva stared as he lumbered slowly to his feet. She looked puzzled, confused that his armor still worked. Jim touched his sleeve and activated the light-shifting properties of the skinsuit. He was invisible to her once again.

"Nice trick. Won't help. You can't touch me," she said and then started to whirl around like a top, arms extended. Jim knew that if she connected he'd probably be killed.

He had one option. Touching the small disc on his jaw, he linked to Marta and Dana and subvocalized, "Querida, I love you. Dana, I love you. Take care of your mother." Then Jim crouched and exploded forward, catching Eva at the knees, below her whirling arms. She was small and he lifted her off her feet. The momentum of his charge hurled both of them towards the windows on the other side of the workspace.

The collision rocked them both, but the window's nanoglass held. A second later she started to beat him about his back. His skinsuit kept him alive, but barely conscious. Although each blow was cushioned, he could feel himself weaken. With a scream, Eva drew all of her strength and reached back to strike a killing blow. Her fist hit the window behind them with a strength that was amplified by her madness and the exoskeleton. It was enough to crack the glass.

But the window held.

"What's he doing?" Marta asked in a quavering voice. "Did that sound like a goodbye to you?"

"Mom. You've got to trust Dad. He'll be okay. We have to find Eva's key."

Dana paused and peered intently at his mother. "Mom? Maybe you should sit down. You look pale and your eyes are red. Is it MAS? Mom?"

"I'll be okay. I'm just going to sit a moment."

She felt tears starting to stream down her face. She wiped her cheeks. The tears were bright pink. She said, "You're right. Let's keep looking."

"Mom...if you're having an attack, you need to rest. Mom? Mom?"

Eva's face was placid. *This is it*, Jim thought. *I'd hoped we'd go through the glass.* He'd pinned her to the window but knew he couldn't hold out for long. She began to beat him. Steady, methodical blows rained down on his back. The pain was excruciating, and his skinsuit armor transferred the impact to cover every inch of his body. With a last effort, he reached out to grab her arms but her amplified strength overwhelmed him. He was looking down at Eva's teak wood floor and watched with detached interest as his field of vision began to narrow.

Eva had won.

Dana scanned Eva's office. Her aerie was barren. A desk and chair. Pillar. Carpets. The standard wall decorations: diplomas, photographs of Eva, and the scarab brooch.

Dana stared at the framed bauble. "This thing bothers me, but I can't figure out why," he said. He took the brooch down from the wall and out of its frame as he had several times in the past hour.

"Mom. I have an idea. I need a nanoscale microscope, something with a resolution down to say, five or ten nanometers. Is there one in your workspace?"

Marta didn't seem to hear. Her face was ashen, with streaks of blood-stained tears.

"Mom? Are you okay?"

Her mouth moved but no sound emerged. Her breath hitched.

"Mom, what's wrong?"

❁   ❁   ❁

Pain shot through Jim's body. Eva struck with machine-like regularity. His vision was reduced to a tiny patch of the floor below him. He noticed the fine grain of the wood and tried to conjure up an image of Marta and Dana. He didn't want his last thoughts to be of building materials.

For a moment, Eva stopped. She bent her head down and in an low voice, nearly a whisper, she cackled. Her speech was rough and accented, as if she'd still been in Sofia. "It was simple, Jimmy boy. I knew Rockford to fail. I examine and I see. Design was not good enough. But that is not my problem. Then they blame me? I just say, 'I quit.' All I had to do. You should have believed."

She stopped for a moment and cradled his head in her arms. "You were my friend, Jim. Then you treat me like a freak. You don't talk to me. You look the other way."

"Now I hold you last time. I wish I could see you now. Turn off skinsuit."

The effort even to subvocalize was now beyond Jim. He stayed invisible.

"No? You stay hidden? I love you anyway. But you hurt me. You hurt me much."

She raised her fist for a killing blow.

Then a shout erupted from the door to Eva's workspace, a cry of animal pain and rage. Struggling to keep conscious, and with a slow, agonized effort, Jim turned his head. For just a moment, his vision returned. There was Rafael. He was out of captivity, still wearing the security collar. He was alive, enraged, and in agony. With a roar the man hurled himself at Eva, arms outstretched as if to embrace her.

Rafael wore neither exoskeleton nor armor. His strength was fueled only by his anger. It was enough. On impact, the weakened window disintegrated, sending shards of glass and three bodies hurtling through the fourth-floor window. They seemed suspended for a moment and then plummeted to the street below.

The window from which they were propelled by Rafael's charge was typical of nineteenth-century construction. Each floor featured 14-foot ceilings. The window was 56 feet above the pavement. Eva struck headfirst with a force that exceeded the limits of her exoskeleton's strength. She was dead even as the other bodies hit the ground.

Jim's skinsuit was enough to stop a fist, but the combined weight of three falling bodies transmitted too much energy to his punished body. The weakened silicon armor was useless. His heart and lungs were battered and his brain bounced within his cranium, the force far greater than the cushioning effect of the cerebral fluid surrounding it. He was already unconscious as he struck the pavement and then he joined Eva in death's embrace.

Rafael landed on top, capstone on a pyramid of bodies. He might have survived the fall but for the security collar. Although its output was diminished, the combination of its relentless release of microwaves, coupled with the shock of impact, stopped his heart.

Less than a mile away, the executive office of the Boylston Street headquarters of NMech echoed with twin horrified cries. Marta and Dana heard Jim's parting words, a collision, a roar and then the sound of shattering glass. Before they could comprehend the sequence, they heard the whoomp of flesh striking pavement. Then silence.

Marta slumped to the floor. Her breath came in short gasps. "Oh no. No, no, no, no…" Dana came to his mother's side and held her. He opened his mouth and clenched his eyes shut, and uttered a wail of grief as he pulled his mother even tighter. She turned and held the child to her breast. Their anguish was heart-wrenching. NMech employees rushed into Eva's office and found mother and child locked in an agonized embrace.

"What is it? What happened?" one of them cried. "What?"

"Jim…Jim…Jim…" Her words tapered off into inchoate cries of despair.

"What happened?"

Their only response was convulsive sobs.

30

RECOVERY

BOSTON, MASSACHUSETTS

MARCH 4, 2045

Marta clung to Dana like a shipwreaked sailor might cling to a a rock. She turned to him and brushed a lock of hair out of his face and wiped tears from his cheek. She kissed his forehead, sobbed again, and then caught hold of herself. She struggled to regain her composure.

"We have to figure out how Eva started this," she said.

"But there's nothing here. What are we missing?" Dana asked.

Marta's self-control cracked. "You mean besides everything that Eva destroyed? Besides that your father is dead? And probably mine? Other than that?" Now her voice was near hysteria. "If there's some way to stop this disaster, she hid it too well."

Dana's head snapped up in sudden realization. "Hidden? Mom,

I think I can find the key to Eva's programming. Something she told me a long time ago about hiding things in plain sight. Come on, we've got to do this and then we can, well, whatever. Where is there a nanoscale microscope?"

Marta lumbered to her feet. She teetered and fell back. She grabbed for the edge of the desk but missed. She collapsed.

"Mom? Mom? Mom!" Dana reached down and touched the side of her neck. Her pulse was thready, her skin cold and clammy, her breathing shallow. Dana cradled her head in his lap and called out, "Somebody help! We need help! Link to Emergency Services. Please."

Several NMech personnel rushed in and found Marta, prone, legs sprawled open, as if welcoming death as her lover. Dana knelt beside her and stroked her hair and face. His face was a map of fatigue and grief.

"What's wrong?" a woman asked.

"Link to Emergency Services. Now."

"Dana, they're all out on emergency calls. Do you know what's going on out there?"

"Listen," Dana said to the woman. He subvocalized a holo display, inviting her to look. "Do you see what I'm prepared to send you? I'll give you her doctor's cloud data. Get hold of her doctor and get him here *now* or she's going to die. Keep the money, share it with the doctor...whatever. But get medical help while there's still time. Please," he begged.

The woman took in the sum, ready to be transmitted. Her eyes widened for a moment and then fixed on Dana. Her voice was gentle. "Dana, there's nobody to reach. The city is under martial law. Most of the country is. All medical personnel are at hospitals or with ambulances. I would do anything for your mother. But it's impossible."

Dana groaned. His cry built to a banshee's wail.

A researcher at NMech burst through the door, a physician before joining NMech. "What's going on? What happened to Dr. Cruz?"

Dana summarized crisply, "She's thirty-six years old, severe JRA, and having an attack of MAS." Then his voice cracked, "Please help my mom."

"Okay, son. Let's see what we have." His voice was calm. Before Dana could move away, Marta reached with one hand and clutched his wrist. Though weakened, her grasp was enough to hold him fast. Dana bent down and put his ear to her mouth.

"Go...stop Eva. Nanoscope in my workspace. I love you, son, with all my heart, with all my soul, and with all my might."

"Mom, you're going to be all right. Hold on. The doctor's here."

"Dana. Listen to me. You must go to El Yunque. Find Abuela. She'll know what to do."

"Mom, don't talk like that. You're going to be okay."

Dr. Marta Cruz, bohique and researcher, mother and widow, the scientist credited with ending the Great Washout—or helping to start it, depending on the account—summoned her last reserve of strength. "Hijo! Promise me. Whatever happens, you must go to El Yunque. Promise me!"

Tears streamed down his cheeks. He bent down and embraced her. "Oh, Mama. I promise. But you have to promise me that you will live."

Marta smiled. "I promise that I will love you always and my spirit will look after you." She let go of his wrist and reached behind her neck. Clumsy fingers unfastened a string that held a small leather pouch to her breast.

"Dana, take this. You will find someone to wear it. Abuela can teach her, too." Softly now, "Go to Abuela."

Dana stared at his mother's leather pouch. Marta's voice trailed off, unintelligible now, a series of moans. She was semi-conscious. And then, silent.

The doctor pushed Dana out of the way, ripped apart Marta's shirt and applied medical cloth to help regulate her vitals, a vain gesture that would do little more than rob the body of its modesty. Dana turned his eyes away from the sight of his mother's torso. Too cheerful sunlight streamed in through the window, and reflected off the dull surfaces of Eva's furniture. The shadow cast a gray pallor on Marta's slack face. The color of life was gone.

31

MY MOTHER

FROM THE MEMORIES

OF DANA ECCO

Imagine waking up every day with a stiff neck, unable to turn your head to the left or to the right. Imagine your back, legs, arms, hands, and hips, as stiff as a rubber toy left overnight in a snow bank. That was my mother's every morning.

She never complained.

What would you do if your wrists, knees, spine, shoulders, jaw, and ankles were swollen, hot, and tender? Your fingers puffed at each knuckle? Would you cry out? Seek the comfort of human sympathy?

My mother did not complain.

How about the fevers, aches, and fatigue? "Ah," you would say, "That I can bear. I'd force fluids, nip some whiskey, and take to my sickbed for a few days." But what if these symptoms persisted, not for a few days, but for years? Would you beg for mercy? Or take the advice of Job's wife, and "curse God and hope to die"?

My mother did not complain.

What if you bled during times of stress? The odd bit of deep muscle hemorrhage or retinal bleeding? Would you shriek in terror, one fine morning, if your eyes were red-rimmed from blood?

My mother endured all of this silently, cheerfully, even with humor. I remember the year she greeted the neighborhood trick-or-treaters at Halloween red-eyed with blood dripping like tears. Few costumed visitors returned the next year.

At the end, would you accept your progression from morbidity to mortality? Or would you "rage against the dying of the light"?

My mother suffered from a chronic illness, juvenile rheumatoid arthritis, JRA. Macrophage activation syndrome or MAS is a painful and life-threatening side effect of JRA. Microphages, literally, "big eaters", are white blood cells that consume debris and pathogens in the body. If these microphages rampage out of control, they cannibalize the body. MAS's effects are rapid and often fatal. The stress of the Great Washout triggered an MAS episode.

My mother was a healer and a researcher and she had lived with JRA for years. She understood the significance of her symptoms. Had she sought medical treatment immediately, she would not have collapsed on the eve of the Recovery. Instead, she stayed focused on discovering how Eva Rozen triggered the Great Washout.

My mother didn't complain. She merely left this world with one more orphan.

32

CERBERUS (II)

FROM THE MEMORIES

OF DANA ECCO

Years after the Recovery and the humiliation of a lengthy inquest into my parents' role in the Great Washout, my anger is still fresh.

Dr. Luminaria, the behaviorist who mentored my father, explained to me that the unconscious mind lacks a sense of time. Events that made a mark on me years ago are still current affairs. The mind's ability to capture sensory input is unimaginable, but it hoards information, doling out memories with a parsimony that would embarrass a miser.

Another agent works with the same automatism as the unconscious mind. My body colludes with my memories and floods me with the chemistry of emotion—cortisol, adrenaline, acetylcholine, catecholamine. I rage, weep, and cower in equal measures, just as

Eva Rozen raged for the whole of her unhappy life. My conscious thoughts might dwell on the beautiful or the mundane only to be washed by a bath of neurotransmitters offered by the rage of an eternal fifteen-year-old child who dwells within my unconscious mind. In an instant, I may shiver with fear, quake with rage, or drift into a fugue state—then wonder where I'd gone. The world had its recovery. When will I have mine?

❀   ❀   ❀

My mother lay dead in her work area. I bent down and kissed her eyelids and cheeks and lips. I picked up her medicine pouch and Eva's scarab and walked to my mother's lab. I felt numb, a blessed sensation that would pass all too quickly.

I powered the nanoscope. The device sprayed a phased pattern of X-rays above and below its target. The emissions have a wavelength of just over one-tenth nanometer so it was accurate to the atomic level. The nanoscope analyzes diffraction patterns and produces a detailed image of an object's surface and electrical composition.

I focused on the scarab. The nanoscope was maddeningly accurate. It was like searching the boardroom conference table with a jeweler's loupe to find a single grain of salt.

I cursed Eva and her damned scarab, small enough to fit in the palm of my hand, but with enough relative space at nanoscale for the contents of an entire library. Where to look? I remembered Eva's words, "If you want to hide something, put it in plain sight, but make it very, very small" and started with the irregularities in the pin. On the third try, I found her journal. But I faced a bigger challenge. It contained thousands of pages.

I tore myself away from the nanoscope in frustration and helpless rage. How could I find what I needed, what the world so desperately needed, the key for which my parents had given their lives? How would Eva have tagged the information?

I returned to Eva's workspace, averting my eyes as I passed my mother's corpse. The doctor and two NMech admins were tending to the body and looked up at me. Judging by their expression, my absence from her corpse was incomprehensible. I continued before they could try to console me.

Eva's scant possessions were lined up on her desk. An entire lifetime contained in a half-dozen photos, diplomas, and a few pieces of art. I looked again at the photos and artwork. Nothing there. I was running out of time. Where would she have hidden the key I needed?

Then I remembered my last interchange with Eva, when she penned me in an unlocked cage and instructed me to jack nearly a hundred datapillar accounts. One of the accounts bore no name. It had only one item, a strange piece of artwork. At the time, I gave it little heed; events were starting to move too quickly. I had assumed that the unnamed account was hers.

Now the item called out to me. I invoked a heads-up display and looked at the piece contained in that anonymous account. The image was of an antique lithograph, out of place in the ultra-modern sterility of Eva's world. It portrayed a powerful three-headed dog, eyes bulging, mouths snarling and snapping. One massive paw clutched a human figure. In the lower left corner was a word, all caps: "HELL", and below that, "Canto 6." Why would Eva have it? She hated dogs.

I scanned the print with my sleeve and had the office pillar search for the print. In less than a second, I had the answer that had eluded me for hours.

❀　　❀　　❀

My mother was still lying where she collapsed. Her colleagues stared at me as I passed by, heading back to the nanoscope. I had no time to stop. I'd found the key. The lithograph was by William Blake and it portrayed Cerberus, the guardian beast of the underworld. It fit Eva's sensibilities perfectly.

Cerberus. That had to be the password. I moved her private journal from her pillar to my sleeve, invoked a display, and skimmed her notes. They were clear, precise, terrifying. The enormity of what she had done made me reel. First, the test cases. A water desalinization plant, disabled. A squad of UN soldiers rendered helpless, overrun. Then kidney dialysis, insulin regulators, terminated for hundreds of thousands, and medication ended for millions by a simple electronic command. What good is a miracle when it is controlled by a madwoman?

Eva's attack was indeed launched from her home pillar. I ran for the street, ignoring the calls of those attending to my mother's body. I would grieve later, but now I need to get to Eva's home as fast as possible. People were dying and I had no time to spare.

I looked across Boylston Street toward Commonwealth Avenue. Four blocks to Eva's home. An easy jog. But it would be surrounded by emergency workers. I doubled back into my mother's office and found an NMech military-grade skinsuit with cloaking capabilities. I donned the suit and ran back across Boylston Street and the short distance to Commonwealth Avenue. I turned left and headed west, one long block to Clarendon Street and then a few feet further to Eva Rozen's home. Six minutes had elapsed.

I was able to avoid police and emergency workers, but not the view of a pyramid of bodies. It was an angry canker on the street. My heart lurched and my gorge rose. I turned away, and scanned the front of the building. There was the fourth-floor window, my father's passageway to the concrete below and my destination. I slipped in the front door and hurried up the stairs to Eva's work-space. There was a pillar. I was gambling that it was the one that had launched Great Washout.

I approached it cautiously, scanned with my sleeve and found a data sensor. I triggered a burst from my sleeve, a software cue. The pillar demanded a recognition code. My sleeve emitted a single word, Cerberus. The pillar's status light turned green. My sleeve pinged.

I was in.

I scanned the programming, afraid that gaining access was the easy part, that Eva's programming would be incomprehensible. But she was an economical coder, well-organized. She had created an elegant application. It was exactly as Denise Warren had described: there was a sub-routine in the accounts receivable programs that had shut down customer accounts for non-payment. This was an outcome for which Denise had prepared me. I pointed my sleeve at Cerberus and another data burst travelled to Eva's pillar and deleted the rogue code. Each account's payment status changed to current. It would take several seconds for all of the accounts to reset, and it would be too late for hundreds of thousands of victims. But millions of others would live.

❁   ❁   ❁

Emery Miller, Sergeant Mike Imfeld, Nancy Kiley, and Jagen Cater, may they rest in peace. Kiley's staff survived because the threat posed by region-wide rioting ended when their desal filters came back on line. Kidney patients regained their bearings and backed away from renal failure. Diabetics found their insulin levels returning to normal.

My parents were dead; Colleen Lowell was dead; the Eva Rozen I knew was dead and so was her *doppelganger*. I was alone.

33

DEAD MAN'S SWITCH

BOSTON, MASSACHUSETTS

MARCH 4-7, 2045

Twenty-one point eight seconds after Eva Rozen plunged to her death from her Boston brownstone, an electronic Presence awoke. It had been programmed to lie dormant unless a signal from Eva's datasleeve ended. It was a dead man's switch, triggered by Eva's death.

The Presence, a sub-routine within Eva's home datapillar, reached out with electronic senses. It noted human biological signatures in the Rozen mansion. Immediately, it returned to dormancy. The detection cycle went unnoticed, lasting a mere two milliseconds. The Presence repeated its cycle of animation, search, and dormancy until there were no indications of any complex organic life forms in the dwelling, some three days later.

Finding itself alone for the four seconds required to carry out preprogrammed instructions, the Presence sent hundreds of data-burst signals. Most of these went to financial institutions around the world. Three found human targets and sent pulses to their datasleeves.

One found the Governor of the Commonwealth of Massachusetts, slipped through her sleeve's security and pinged an urgent message. The second reached a newly-appointed Special Prosecutor. He was in a press conference and would not see the message for seven minutes, during which time all hell would break loose in the governor's office.

The last signal activated software that had been downloaded days earlier, when Eva Rozen pushed gently, one final time, on Dana Ecco's forearm. Dana's sleeve accepted this last inbound transmission without notifying the preoccupied scion of the Cruz-Ecco family.

34

 SPECIAL PROSECUTOR

Suffolk County District Attorney Sean Doyle, elected to the first of the public offices he coveted, had progressed in the years since Jim Ecco's trial for assault and subsequent conviction for disorderly conduct. Doyle rose steadily through the legal system and he won the DA's office by a comfortable margin two years earlier. Now the legal and political powers that controlled the Commonwealth of Massachusetts believed that the best choice for a Special Prosecutor in the matter of the Great Washout was Sean Doyle.

Granted, there was the small matter of determining exactly whom to prosecute, but Sean Doyle would be the People's Champion once again.

Doyle kept his trademark navy pinstriped suit, changing only the material to a lustrous nano-silk befitting his enhanced station in life. His red and blue striped club tie still formed itself into a perfect Windsor knot which never loosened or came askew. He no longer needed enhancements: the gray hairs that salted his blond curls fit perfectly with his image of energetic maturity. He strode with purpose into the State House and walked to a podium to greet the media. The event was important enough that the reporters attended it in person, rather than in virtuality.

"Ladies and gentlemen, I have a brief statement. First, let me thank all of the emergency personnel who helped avert an even greater catastrophe than we might have suffered. My staff is working with governments around the world to ensure that water and medical supplies will not be interrupted again. We restored service to NMech customers just hours ago. Now it is time for an accounting and I can assure the public that we are doing everything in our power to bring those responsible for this vicious act of terrorism to justice."

"Three persons of interest were killed during what is being called the Great Washout. We believe that they may have had knowledge of how this catastrophe was committed. We have pledged every resource to learning exactly what happened so that we can prevent another attack. The combined resources of the City of Boston, the Commonwealth of Massachusetts and the United States government have been placed at my disposal to ferret out the truth and take appropriate action."

"I will not take questions today. Again, let me thank the emergency responders who prevented a much worse tragedy, and the valiant efforts of those, who, under my direction, stopped the Great Washout. We have begun to restore normalcy to the world."

35

 GRAY GOD

**FROM THE MEMORIES
OF DANA ECCO**

The family home was now mine although keeping it would prove to be a challenge. My parents' estate, including the house and their NMech stock had been held in a trust. Their wealth was to transfer to me upon their deaths. On paper, I was one of the richest people in the world.

A battle over my inheritance had already begun. Sean Doyle, Governor Azevedo, the mayor of Boston, Congress, and an army of attorneys were seeking to pry it all away. I was too young, immature, too vulnerable to inherit property. A custodian should hold and manage my wealth until I was twenty-one.

Never mind that my education in wealth-management was more thorough than the self-appointed guardians' knowledge. They wanted my parents' estate and would find a way to wrest it from me. Just being in control, even for a few years, would be lucrative. And the possibility of carving up the estate for their own purposes made the prospect of a bruising fight worthwhile.

Eva's share of the company was in dispute. The City of Boston lay claim to it as did the Commonwealth of Massachusetts and several agencies of the United States government. Sean Doyle assigned himself the ambitious task of trying to find a legal precedent by which he could appoint himself executor of Eva's estate. Then there were the lawsuits. It would take years to probate her holdings and to settle with the millions who suffered at her hand—a legal limbo that would prove frustrating to all except to the attorneys who poured their best billable efforts into tangling and untangling Eva's affairs.

The ersatz beneficiaries did not fail to consider that the portion of Eva's holdings and my parents' estate included large blocks of NMech stock. If the company failed, the value of the estates would be slashed. They had a vested interest both in keeping NMech alive and in appropriating it.

I focused on my parents' legacy, not the estate, on their memory, not their money. I used my ghosting skills to have their bodies released from custody then slipped electronically into the pillar of a funeral home and had their remains picked up and delivered to our home. Our home? My home. Now they lay interred near the tree-lined edge of a pond where the Muddy River trailed one last streak of wildness within a great city.

I tried to meditate on their lives but I could not be still. I paced relentlessly, as if expending energy could ease the ache in my heart. Instead, I grew angrier with each step as the shock wore off and rage surged into its place, like a tidal comber filling a rocky void.

I looked through the living room's floor-to-ceiling windows

and contemplated my parents' graves—unmarked save for a spray of the healing flora my mother cherished. Soon the plants would wither and freeze in the gloomy Boston winter.

I tried to convince myself that there was a future. I imagined the view from the living room windows in the seasons to come. The earth would celebrate my parents' sacrifice in three seasons: pastoral spring, teeming with birds, delicate flowers, and tender buds dotting the trees as if from a pointillist's brush, a reanimated totem of hope; summer's heavy green blanket, marked by slashes of floral color—yellow asters, orange day lilies, multi-hued clematis—an impressionist rendition of lustful nature; and fall's crisp cool, concentrating sugars and pigments in the foliage, a harvest celebration of dappled yellow and purple and crimson, a multihued expressionist cry of mortal beauty.

It was a beautiful future, until winter, Eva's season, an inanimate still life.

Today, spring hid. It waited to heave up through the rime. My anger would not wait. Every muscle in my body was tense. I paced and paused, looking out at the graves as if scanning for reanimation, for resurrection.

My parents died for a world that cared little for their deaths, one that tarred their sacrifice with questions of complicity, grave inquiries by solemn pundits. They were being *investigated*. Was my mother an accomplice? Had she been guilty of careless science? Did my father join Eva in death as a foe or as an ally? Why was my grandfather, a convicted terrorist on the scene? Many took that as proof of my parents' involvement in the Great Washout.

As the enormity of what had happened became clearer to me, my tread turned heavier. I could feel my heart break with each step. I replayed my father's final words to me, "Take care of your mother." I remember first thinking that I had failed, then realizing that my opportunity to care for her had been stolen.

Now the survivors demanded an inquiry. No emotion but rage would suffice. I would collapse without it.

At last, I truly understood Eva.

The next several hours are still unclear to me. I remember that I screamed and sobbed, cursed God and then tried to negotiate with Him. Eventually I slept.

Sleep is the great palliative and I awoke with an appetite for food and for justice. I pored through Eva's journals and drew three conclusions. She was brilliant; she had become a different person. And she had planned a catastrophe: gray goo.

Eva had used the puzzling phrase repeatedly in her last several entries. I invoked a heads-up display and found that in 1986, a nano-technology pioneer considered the possibility of a nano-induced apocalypse. He envisioned trillions of self-replicating nanobots that would break down the structure of all matter on earth into what he dubbed grey goo. The vision was chilling. It was also demonstrably unfeasible. By 2004, the futurist withdrew his apocalyptic hypothesis and confessed, "I wish that I had never used the term."

Nano-scientists considered the matter closed. Eva didn't. Nanoapplications need not turn the earth into mush to suit the thing she had become. There were other solutions. Bind even a little of the atmosphere's oxygen to carbon and the resulting carbon dioxide would kill us all. Develop an artificial microphage, one that attacked cell membranes like the rampaging cells that killed my mother. That would render humans into a primordial soup, while leaving the earth whole.

Her notions were farfetched, I thought. It would take decades to achieve a toxic mass of CO_2. She would have to protect herself from the marauding microphages. But if anyone could do it, it would have been Eva. She was a genius. Her reason had been destroyed, but not her skill.

Eva left me a second puzzle, the last entry in her journal. At first, it was incomprehensible. Her voice must have been jittery and the transcribed words were broken. I invoked an edit program and reorganized the characters into something meaningful. Her last words were in Latin: "Nemo me impune lacessit," and then the initials, "EAP."

I linked to the home pillar and came up with a translation and source of the Latin phrase. It was used by the Most Ancient and Most Noble Order of the Thistle, a Scottish chivalric order created by King James VII in 1687. It is the motto of the Royal Coat of Arms of the Kingdom of Scotland. It was used by law enforcement agencies on the mourning bands commemorating a fallen officer. It is also the family motto of a murder victim in an Edgar Allen Poe short story, "A Cask of Amontillado."

The motto's translation was a fitting epitaph for Eva Rozen. "No one insults me with impunity" or, "Touch me not without hurt."

Of all the things that confounded me in those days, this was the most puzzling. I could not imagine Eva reading 'worthless stories.' Yet she must have in order to have quoted Poe.

But that was not the end of Eva's legacy. She left me a final gift. It was waiting for me on my datasleeve.

36

A PROMISE

FROM THE MEMORIES

OF DANA ECCO

Dr. Luminaria explained to me that anger is a normal reaction to loss. She compared anger to a life preserver. It is an emotion strong enough to keep a person from drowning in sorrow after a tragedy. Once on dry land, a healthy person discards the life preserver. But many cling to the anger as their misfortune continues to play as a current event in the unconscious. My anger would remain with me for many years and play a near-fatal role in my future.

❀   ❀   ❀

After I disabled Cerberus, I returned to NMech, to my mother's workspace. It was hard to see. My eyes brimmed with tears. I sank into the deep pile carpeting my mother favored—an easier surface for her diseased joints to bear. I approached her pillar and invoked a program on my sleeve. I had determined that neither her work nor Eva's would be left behind. I alone would decide how the world was to benefit from my mother's labor and I could not allow Eva's notes into anyone else's hands. Good and evil would be in my hands. Yocahu and Juricán.

I used my sleeve to relay my mother's research to my own private pillar. I had done the same at Eva's home. I reset both pillars to an electronic *tabula rasa*. 'Swipe and wipe' Eva called it.

I could develop the work of the erstwhile colleagues—Eva's weaponry, my mother's medicines. Either would take time. I would have to immerse myself in the study of medicine and biochemistry. But I believed I had time, and I'd been trained in science by experts of no less rank than Eva Rozen and my mother.

I took a final look at the place my mother spent so many of her hours, the inviting salon that had been a second home to my father, who enjoyed the simple act of watching his wife lose herself in her work.

It was all gone.

One keepsake caught my attention—an old-fashioned photograph. It was of my great-grandmother, Abuela. I remembered my mother's tales of Abuela's great healing prowess. "She saved me," my mother had said, recounting a childhood summer in the rainforest. I picked up the photo and looked at the lined face of the old woman. Was she really a shaman, a medicine woman? Was the lush green canopy behind her, El Yunque, really a place of magic? Enough wool-gathering. There would be time later for contemplation.

As I set the photograph of Abuela back on the workbench, a memory beckoned. "*Hijo. You must promise me that whatever happens, you will go to El Yunque. Promise me!*"

I had not paused long enough in the last few days to respond to my mother's demand. But the voice was insistent, and I had promised my mother. I would go to El Yunque.

37

MEA CULPA

Sean Doyle was summoned by state capital security officers to appear 'forthwith' at Governor Mariana Azevedo's home near the State House on Beacon Hill. Azevedo wished to avoid the press, and the Massachusetts Constitution does not provide an executive residence for the governor, although Part the Second, Chapter II, Section I, Article I, of that document invests the state executive with the title, "His Majesty." Azevedo briefly considered reinstating the title after her election but considered the gender reference to be a demotion.

A hand at each of Doyle's elbows guided him gently but firmly to the governor's home, six blocks from the press room at which

Doyle had been speaking. The streets were cleared of P-cabs and private cars, and they arrived in minutes.

"Why are we meeting here?" Doyle blurted out when he stood in front of her desk.

Azevedo wore her trademark white baiana dress and turban, the traditional garb of the Bahia region of Brazil. Brazilian immigrants to Massachusetts had become the second-largest voting bloc in the state, and the occupant of the governor's seat was likely to be determined by the Carioca and Baiano population for many elections to come.

"Sit down, Sean," Azevedo ordered. "What are we going to do with Eva Rozen's message?"

"What's to decide? A full holographic confession? Details on how she triggered the Washout? Where's the issue?"

"It's perfect for you as Special Prosecutor, but what will it do for the Commonwealth?"

"You mean for your reelection," Doyle replied.

"Sean, are you planning to run in '48?"

"Well, a confession, a conviction of a mass murderer, etc., etc., Madame Governor—I'm not going to waste that kind of political capital."

"For chrissakes, Sean. Drop the formalities. Do you see any cameras in here?" Doyle looked around and then shook his head. "Then let's not dance around this. If we release Rozen's message, how do we both manage to get what we both want? You've been drooling over my office for years, and I just fought tooth-and-nail to win it. What will you take instead?"

"Why do I have to take anything if I have her confession?"

"Simple. I fire you on the spot for malfeasance and appoint one of my own people as Special Prosecutor. A few years ago, this nobody, this—what's his name?—this Jim Ecco character gets off with a disorderly instead of a felonious assault. You screwed up the prosecution, Sean. Now he lands in a pile of bodies with the worst mass murderer in U.S. history. How's that going to play to the voters?"

"I did no such thing. It was a routine plea bargain." Doyle's pale Irish features reddened. He shouted, "And that was years ago!"

"Calm down, Sean, or you'll bust a blood vessel and trigger a med-alert. That won't play well with voters."

Azevedo watched as Doyle grew still. He looked dangerous. When he spoke again, she heard control return to his voice. "Attorney General in '48," he said, "and I want your support in '52 if I run for governor or for the senate."

"I'll support you for AG. Unless you screw this up or cross me, you'll pretty much run unopposed. You can have the AG's office in '48 but you support me for the senate. You can look at a senate seat after a couple terms as governor—hell, you could try for the Oval Office then. But I announce Rozen's statement. Deal?" asked Azevedo.

Doyle rose, his pinstriped blue suit following him as carefully as a diligent mother of a two-year-old child. His club tie remained perfectly knotted. He offered his hand. "We have a duty to the people. You announce and then I'll take questions."

Azevedo winced inwardly at Doyle's pompous rhetoric, then smiled. Her political future was secure, at least for the next eight years. After that? Well, she just might see Doyle again, likely on the hustings during a presidential primary race some years from this day.

Azevedo's flowing white dress and Doyle's pinstripes made an unlikely diptych as the two politicians addressed the press. As agreed, she announced and played Eva's confession and then handed Doyle to the media.

Eva Rozen had recorded a holograph. She stood life-size, four feet, four inches tall, wearing her trademark black cargo pants and a black work shirt. Her hands trembled as she spoke. Her accent, gone since childhood, had returned.

"I have nothing to do with Rockford. You want to know how that happen? Go to Texas to find out. Look at results of tests. Look at containment building. Data is no good. I warn you and you ignore me. You cheer when they finish ahead of schedule. You know how they finish early? Sloppy science. That's why building leaks and explodes. Okay, you pay for that in blood. But then you accuse me of murder? You say I trigger Rockford?"

"Nobody accuse me. You are fools. You will pay. I do not attack you. But I stop my charities. Nobody make me do them. I do myself, I pay for myself, and I stop them myself. You call it public health. Except I pay, not the public. If I cancel anybody by mistake, don't worry. I give refund. All is fair. Nemo me impune lacessit."

The conference erupted. Doyle took his time fielding questions, while Governor Azevedo looked on, looking solemn for the vidbots, and left when the obvious question, "Why did she make a confession?" caught the Special Prosecutor by surprise. "I think she was bragging," Doyle managed, "It was her way of going out with a bang."

In fact, her confession exonerated Marta and Jim. Doyle did not consider that loyalty might be a part of a mass murderer's emotional inventory.

A reporter cornered Azevedo backstage and asked how she felt. She peered at the reporter and noted the presence of vidbots. "How do I feel? Terrible! Thousands of people died. However, I am satisfied because a killer will be brought to justice, if posthumously. Her presumed accomplices were cleared of wrongdoing. In fact, I am going to issue a proclamation honoring the memory of Dr. Maria Cruz, who was a hero."

"What about her husband, Jim Ecco?" the reporter asked.

"Yes, him too."

❀ ❀ ❀

Dana Ecco's datasleeve was the third human target of the kill switch's transmissions. It activated a series of commands that had lain dormant in the sleeve. There followed hundreds of electronic conversations. These flashed from Dana's sleeve to financial institutions targeted earlier by Eva Rozen. Data sped back to his sleeve. One by one, Eva Rozen's assets were transferred to her only beneficiary, Dana Rafael Ecco, along with additional software that prevented the transfers from being traced. By the time the claims and counterclaims among Doyle, Governor Azevedo and the United States government were resolved, there would be no financial assets remaining over which to bicker.

38

EL YUNQUE

FROM THE MEMORIES

OF DANA ECCO,

The journey to Puerto Rico had been fraught with reminders of the past days' horrors. I flew in the same NMech jet that had delivered my grandfather to Eva and Nancy Kiley to her death. Sean Doyle had seized the craft, but with the assistance of a handful of attorneys, and a bit of ghosting on my part, I'd regained the use of some of NMech's assets.

I landed in Luis Muñoz Marín International Airport and dismissed the waiting NMech driver and security. The fewer reminders of NMech the better. I stood wilting in the tropical heat outside the terminal, unsure of my next destination. In the end, I took a P-cab, still undecided. I sat in the driverless car and hesitated over several preset destinations. I did not choose El Yunque, but the fortress of El Morro, instead.

The Spanish conquistadores built the stone fort at the tip of the island. For centuries it was an impregnable stronghold on the land they'd conquered. But it was no match for the United States' military might. In 1898, the fort yielded during a brief assault. Soon the defenders capitulated and Puerto Rico became property of the United States. The conquerors called the military action the Spanish-American War. The island population came to know it as the Invasion of 1898.

I walked the ramparts of El Morro's stone walls and the wind asked me, *What is there here for you? Leave this monument to war and find peace in the forest.* Still, I walked. The open ocean hugged one side of the battlements, the old city the other. Both vistas called me—the surging ocean with its wild currents and the well-constructed old city. Would I find peace in wildness or in order? Revenge for my parents' death or healing for the needy?

Night fell, as sudden as a sneeze this close to the equator, and the stress of the last several days caught up to me. I sat heavily in the courtyard of an apartment building. It was high above *La Perla*, the city slums below Old San Juan. I leaned against a wall, listening to the faint sounds of the shanties and alleys below. Sea breezes began a steady march inland and cooled the earth. Stars rose and fell and I relaxed into the rhythms of the night.

The sounds of explosive breathing—a grunting family of pigs—drew me out of my reverie. I lay very still as the sow and her shoats snuffed at my legs. Indifferent, they moved on, heading on their rounds before returning to *La Perla*. It was possible that I was the wealthiest person in the world, and yet I had nothing of interest to a drove of swine.

With that observation, I was ready to travel to El Yunque, to meet Abuela. How would I find my mother's family—my family? I had only a small photograph of the old woman. I could imagine my mother telling me that Yocahu would show the way.

As the sun inched above the horizon, my P-cab rolled to a silent stop at El Portal, the visitor center at the rainforest entrance. This was part of the journey that a frightened thirteen-year-old girl had taken. She had just lost her mother. I'd just lost mine.

I decided to explore the rainforest that my mother held so dearly. Perhaps I'd find Denise Warren, the NMech bookkeeper we'd met at the beginning of the Great Washout. I put that thought aside; it would be another NMech reminder. I wanted to see El Yunque with the same clear eyes that my mother would have brought to bear on it.

Which way to go? El Yunque Peak beckoned, a bristling shard of rock, green-carpeted and crowned with misty clouds. Was this the home of the legendary Yocahu? A place of magic or mere volcanic debris hoisted up during the formation of the Caribbean tectonic plate?

The morning's hike brought me six kilometers to the peak. Would the All-Powerful deity descend from the mountain like Moses carrying the Law writ in stone? What a waste! At least I had kept my promise and come to the mountain.

I sensed the presence behind me before I saw her. No footsteps announced her approach. A small, wizened figure stood near, just as she'd once stood near my mother. Again, she was still, save a crooning voice. The old woman's face was even more deeply lined with sun and age and care but her eyes shone clearly. She spoke the same words to me as she had to my mother.

"*Hijo*" she intoned, stretching out the vowels—child. A single word carrying eight decades of love and wisdom. Abuela. My great-grandmother. She was real.

"*Hijo…Mira aquí.*" Look here. Abuela touched her hand to my heart. "*Estás tan airado.*" You are so angry.

"I'm not angry, Abuela," I had said. "I'm just tired and sad and I miss my mother and father."

Abuela merely pointed to my tightly-clenched fist. Then she took my hand, uncurled the fingers and led me into the rainforest, into a place where I could choose peace and life.

THE END

ACKNOWLEDGEMENTS

Many hands make lighter work and several pair contributed to *Little Deadly Things*. First, infinite thanks to Jody for insisting that a quarter-century's germination was enough. She went all in on my dream. Jody read more drafts than you'll find in an old New England barn. Lordy, I love that woman.

My thanks to:

My father for inspiring me to write. He compelled me to write compositions on Saturday mornings, on a light blue, three-legged stool in the bathroom. I wished I'd asked him why he chose that particular venue.

Dr. Allison Lloyd McDonough, medicine woman and healer, who answered more medical questions than appear on the MCAT exam.

It's easy to create a villain, to dream up dirty deeds. It's not so easy to imbue a villain with humanity. I would have failed without expert guidance from Dr. Steven Krugman and Sybil Houlding. Neither allowed me to remain in the flat-plane world of Bad Seed cliché. I struggled for four years to bring Eva from the unformed to the expressible and would not have found her character without Sybil and Steve's astute coaching. Sybil is the brightest living light of the psychoanalytic community. That she is my sister is irrelevant.

Murray Steinman, P.R. and branding maven. That he is my brother is not irrelevant in the least. Need a good PR firm? Talk to Murray.

Roger Gefvert—one of the two gentlest people I know. He is artful, thoughtful, and talented like you wouldn't believe, and generous with his labor. Thank you, Roger, for the cover and interior design. Wow!

Jordan Rich, for your unexpected generosity in promoting *Little Deadly Things*. You helped me get author blurbs and put me on the airwaves. Cool!

Rebecca Houlding, who helped keep the courtroom scene tethered to reality. She and her father are the world's greatest attorneys. That she is my niece is irrelevant.

Attorney David Ceruolo, for quick answers to many legal questions.

Dr. Lindsay Drennan-Harris, for explaining the chemistry of iron, the Little Atom That Could.

Terri Bright, soon to be *Dr.* Terri Bright. She is one of the country's unheralded experts on canine behavior. She is my mentor and friend, and the only Agility competitor brave enough to run a bull terrier and skilled enough to win. Her training regimen at the MSPCA is unique. My understanding of behaviorism is a tribute to Terri's patient reinforcement.

Crystal Campbell, Keyboard Wizard, for creating a most wonderful website.

Those kind souls who labored through a dismal first draft: Pete Solomon, town librarian and the first to read and encourage; very long-time friends and very dear ones: Jeremy Landau, Melody Urso, and Erin Prophet (tied for first place on the Gentle Scale). Encouragement also came from Lawrence and Elizabeth MacDonald, and from Val Rounds.

Dr. Nancy Krieger, childhood friend, for laboring through a later draft. You are way supportive, and too cool for school, just as you were when we were kids.

Old friend, Jerry Posner, just for being you. Mr. Positive.

Writer and editor Guy Maynard, for his honest reactions and deservedly scant praise of that first draft. Thank you for encouraging me, even though this genre is not your cup of meat.

John Guerrasio, actor and friend. Another who egged me on. "Write! It's good for you!" he said. I admire John because he set out to do what he loves and succeeded.

Eric Sunden, Mr. Patience, for your help with the Kickstarter video…and for pointing out that Fajardo is in the north, not the south of Borinquen. Ooops.

Justin Cattachio, for the Dunedin Kickstarter videos. Oh, heck, all the supporters in that wonderful town.

W. Javier Colon, for help with colloquial Puerto Rican Spanish, and for being young, and idealistic. Be bold.

Michelle Toth, a fine author and teacher, for the inspiration to, and the roadmap for self-publication. Plus, answering a million questions later on.

Amy MacKinnon, another fine author and teacher, who possessed the patience to explain basic writing craft without which I should not have begun and would not have finished.

Editor Stuart Horwitz, of Book Architecture. Professionals make it look easy.

Victoria Wright, Editor Ninja Supreme, for holding me accountable and for being so damned smart. A joy to work with you.

To the selfless and talented folk at Grubstreet. Love, not money, drives them.

To you all, for spending your time with *Little Deadly Things*. Mil gracias.